Conoco:
125 Years
of Energy

CONOCO

125 Years of Energy

Russ Banham

Greenwich Publishing Group, Inc.
Lyme, Connecticut

© 2000 Conoco Inc. All rights reserved.

Printed and bound in the United States of America. No part of this publication may be reproduced or transmitted in any form or by any means, electronic or mechanical, including photocopying, recording or any information storage and retrieval system now known or to be invented, without permission in writing from Conoco Inc., P.O. Box 2197, Houston, TX 77252-2197, except by a reviewer who wishes to quote brief passages in connection with a review written for inclusion in a magazine, newspaper or broadcast.

Produced and published by Greenwich Publishing Group, Inc.
Lyme, Connecticut

Design by Clare Cunningham Graphic Design
Essex, Connecticut

Separation and film assembly by Scan Communications Group, Inc.

Library of Congress Catalog Card Number: 99-68222

ISBN: 0-944641-38-5

First Printing: January 2000

10 9 8 7 6 5 4 3 2 1

The following are common law, pending and registered trademarks of Conoco Inc. and its subsidiaries, either in the U.S. or outside the U.S.:

Benzin, Breakplace, CDR, Coffeebreak and Design, Conoco, Conoco (stylized), Conoco Hydroclear, Conoco in Capsule, Conoco in Thai, Conoco in Thai Capsule, Conoco on Triangle, Conoco Oval, Conoco CDR, Conoco FasGas, Conoco Super, Diesel, FasGas, Freshbreak and Design, Hottest Brand Going, Hydroclear, Hydroclear Diamond Class, Hydroclear Power-D, Hydroclear the Clear Solution, Jet, Jet and Design, Jiffy, Jiffy Shop, Liquidpower, Seca, Seca Diesel +, Sopi, Super, Super +, Super-Cote, the colours blue & yellow, the colours blue, red & yellow, Thirstbreak and Design and Triangle Design.

Conoco and its subsidiaries have also marketed goods under the following trademarks. Conoco continues to retain the common law rights in the trademarks:

Alfol, Branding Iron Design, Concarb, Continental, Continental Carbon, Douglas, E=Qual, Flow Improver, Germ, Germ Processed, Jet Value, Kayo, Oil-Plating, Powerscrub, Ruff-Cote, Touraide, Travel Assistance, Vibroseis and Western.

Photo Credits:

Cover image, pages 5, 164, 171, 172, 210 appear courtesy of Harald Sund

pages 7, 166, 167, 188 appear courtesy of Indusfoto Ltd.

page 17, right, appears courtesy of the California Historical Society

page 26 appears courtesy of 101 Ranch Oldtimers/Jack Webb

pages 27, right, 28, 43 appear courtesy of Ray and Velma Falconer of The Glass Negative, Ponca City, Oklahoma

page 139 appears courtesy of CORBIS/Owen Franken

pages 140, 179 appear courtesy of CORBIS/Bettmann

pages 180, right, 181, 226, 228, 229 appear courtesy of David Gold

page 211 appears courtesy of Haruko

pages 218, 238 appear courtesy of Greg Smith

page 230 appears courtesy of Williams Photography Ltd.

page 231 appears courtesy of Richard Whittern Fine Art

pages 250-252 appear courtesy of Greg Smith/Saba

page 263 appears courtesy of Mel Nudelman

All other photographs and historical items appear courtesy of Conoco Inc. and its subsidiaries.

Conoco would like to express its sincere thanks to its staff photographers who have documented the company's history over the last 125 years. In particular, Conoco appreciates the efforts of current staff photographers John Graham, Garth Hannum, Larry Jones, Rich Ostrem, Dan Taglia and Paul Waffle. Additional photos were taken by employees Keith Blumert, Tom Buek, Gary Edwards, Liz Williamson and Teresa Wong.

Conoco wishes to acknowledge the following people who contributed memorabilia items for this book: 101 Ranch Oldtimers, Franz Ehrhardt, Sondra Fowler, Wanda Jones, Jack Keathly, Bruce Lawrence, Aliene Macomb, Sue Oldfield, Leland Smith and Mike Stinson.

Chapter One
Parallel Passage 12

Chapter Two
Going Global 68

Chapter Three
Crisis and Competence 124

Chapter Four
Realignment and Refinement 156

Chapter Five
Independence 208

Timeline 268
Index 272

Parallel Passage

The engraving at right, from 1872, shows open oil storage tanks and barrels. In 1859, only 13 years earlier, Colonel Edwin L. Drake had completed the first commercial oil well in the United States, marking the birth of the oil industry. The Titusville, Pennsylvania, well produced only 35 barrels of oil a day, which sold at $40 a barrel. But the refiners who were distilling coal to make lamp oil would soon find they either had to change over to petroleum products or go out of business.

Our story begins with two men.

Although Isaac Elder Blake and Ernest Whitworth Marland never crossed paths, their lives are entwined in the early history of the Continental Oil Company, today's Conoco. Each built a company predicated on the promise of petroleum — one rising in the Rocky Mountains and the other in the rolling Sooner lands of the Southwest. Both amassed great wealth only to lose it all — not that it mattered. For in the bluff and call of oilman's poker, the joy is in the journey. Money slips through fingers and leaves no trace. But oil marks the man.

Isaac Elder Blake was always the young man in a hurry, the boy wonder who earned a teacher's diploma at 16, two years before most people even applied for one. Born in Canada but reared in the States, Blake always thought of himself as an American. His father was a Methodist minister of high principles who died tending the wounded of the Civil War. Blake, however, took a more pecuniary path, heading for Pennsylvania to make his fortune in oil.

By 1869, the 25-year-old Blake was earning $1,000 a day buying and selling crude. But in the adolescent, unpredictable oil business, one day's winnings were the next day's losses. Blake made a disastrous miscalculation, speculating on a large quantity of crude that later had to be sold at a loss, leaving him $40,000 in debt. His pockets empty, he followed the lead of other young, ambitious men and, in 1874, headed west.

He planted his roots in Ogden, Utah, a short distance from Promontory Point, site of the golden spike marking the transcontinental railroad's completion in 1869. Founded in 1850 on high land surrounded by mountains, Ogden was known for the immense number of beavers and muskrats along its streams. When Blake arrived in 1874, it was a thriving community, boasting a drugstore, dry goods emporium, hotel and wagon shops, and a place where the Sharp brothers practiced dentistry "in all its various branches."

Most residents, however, still depended upon whale oil or tallow candles for illumination. Coal oil

Isaac E. Blake founded Continental Oil Company in 1875 to provide petroleum products for Western pioneers. Tenacious and optimistic, he organized Continental after losing $40,000 on a previous oil venture in the East. Such was the "now-rich, now-poor" game of oil in the nineteenth century.

Men at work on a primitive oil derrick seek the fortune of black gold. Lamp oil and kerosene, two derivatives of crude oil, would replace whale oil and candles as sources of illumination. Grease lubricated the axles of the "prairie schooners" carrying Americans to the Western frontier. Even with early production methods, such as the 1860s drilling rig depicted below or the "cable method" pictured at right, it cost only $1 to produce a barrel of crude oil, versus about $35 to produce a barrel of whale oil. The potential profits lured many, but the oil would prove elusive for all but the most daring and intent prospectors.

and kerosene offered far better light with less smoke, but they were extremely expensive. Hauled by bull team from a primitive refinery in Florence, Colorado, kerosene retailed for as much as $5 a gallon. A kerosene lamp in the Wild West was a luxury, a mark of prestige and style.

As the West filled with people, however, Blake saw his opportunity to grow with the oil business. The railroads were pushing back the frontiers, and the wagon trains were rolling. Families bringing with them the amenities of civilization — kerosene lamps included — were replacing the buffalo hunters and mountain men.

Blake conceived the idea of organizing a company to transport kerosene in bulk from the East. The kerosene would be measured into barrels, brought by rail to Ogden and then sold to the pioneers at a cost far below the sealed five-gallon cases then in use. He persuaded four other businessmen to join him in his dream to illuminate the West. Each invested $5,000, and on November 25, 1875, the Continental Oil and Transportation Company was born.

The new company purchased two railroad tank cars — the first in use west of the Missouri River — for the transportation of kerosene from a refinery in

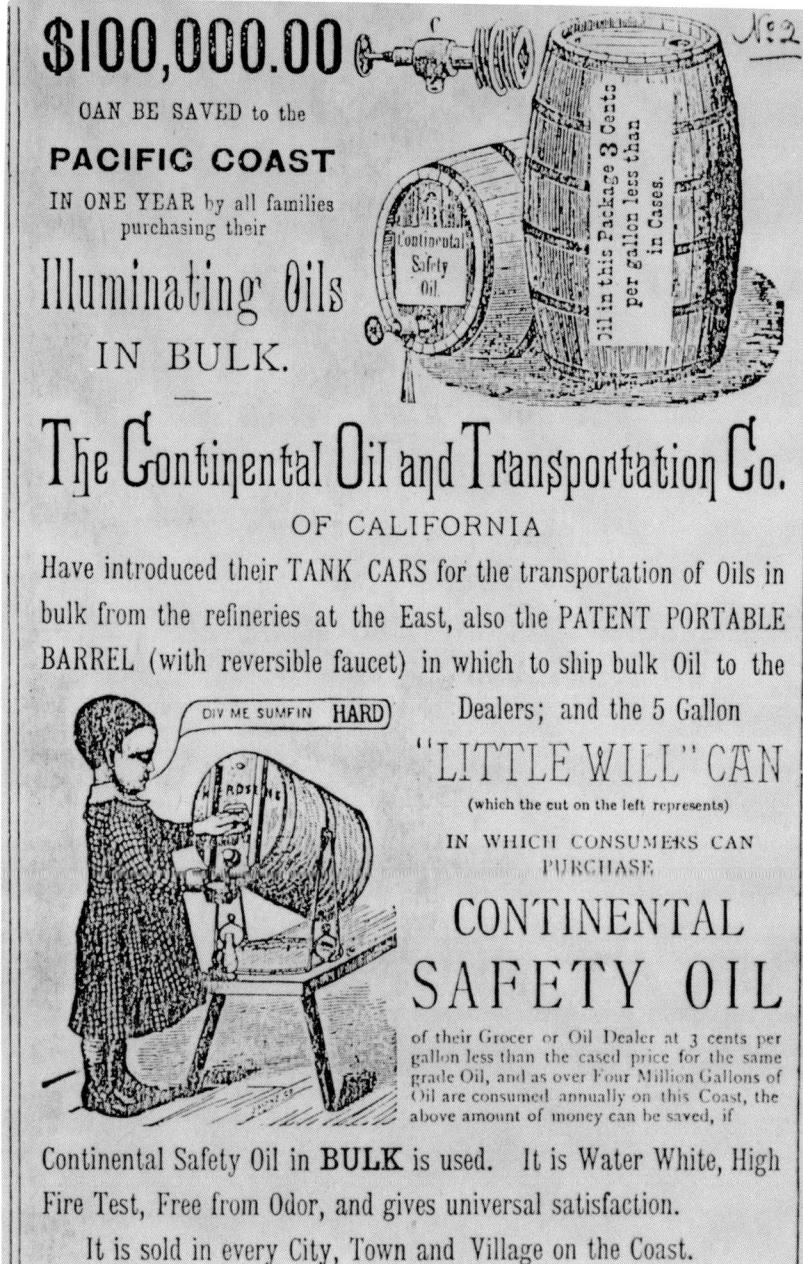

Continental began shipping oil by rail to California in 1877. An early print advertisement touts to West Coast pioneers Continental Oil's patented portable barrel and inventive "Little Will" can, a small five-gallon barrel attached to a rocker that made it easy to pour. Petroleum products did more than illuminate the home. Above left is an ad for a portable gas lantern similar to the type used by miners.

PARALLEL PASSAGE 17

In the early days of the West, horse-drawn wagons transporting barrels of oil served the pioneers filling the West. Early distribution methods utilized warehouses, like this Continental bulk station situated close to rail lines in Butte, Montana, in 1914. Full containers of oil, such as the one shown opposite, were delivered by wagon to customers and then picked up when empty. Grocery stores were the predominant customers at the time, buying a single barrel of oil from which their patrons would extract a gallon or more.

Cleveland, the port city John D. Rockefeller had turned into an oil boomtown. Continental established bulk oil stations at Ogden, Utah; Council Bluffs, Iowa; and Cheyenne, Wyoming, to store the kerosene upon arrival. Sales agents were appointed in the principal cities of Colorado, Wyoming, Utah, Iowa and Montana.

Continental distinguished itself through Blake's unmatched ingenuity. The company became noted for its special steel barrels, encased in wooden staves bound by iron straps. Bungholes were drilled at either end — that way the barrels could be tapped either lying horizontally or standing on end. Continental also introduced a five-gallon oil can attached to a rocker. The rocker let the purchaser pour kerosene from the container without lifting it — the first ergonomically designed oil can!

Kerosene was Continental's main stock in trade but not its only product. The petroleum distillate benzine, used principally to clean stoves, also was sold. Other early products included candles, bulk wax, axle grease and a few non-petroleum products such as lard and ready-mixed paints. Later, the company added harness oil, hoof oil for horses and a popular ointment. W. H. "Billy" Potter, a 13-year-old Continental

PARALLEL PASSAGE   19

An early-day compounding plant in Trinidad, Colorado, is pictured above. The mining activities of the West required illumination. This Continental plant supplied miners with oil for their lamps. At left is a photo from 1909 depicting the company's Missoula, Montana, bulk station. Montana, with its vast copper mines, was also an important marketing territory for Continental. Other prime markets at the time included Wyoming, Utah and Iowa.

office boy in 1898, recalled that the "healing properties of Continental Ointment were so good that some of us used them on our own cuts and burns."

Potter's endorsement had limited value, however, as there were no industry manufacturing standards at the time. "Products from one refinery often bore little resemblance to those from another," company historians reported in 1975. Prices also fluctuated widely and without warning in the budding industry.

With characteristic tenacity, Blake persevered, driving the company into new markets. The Continental Oil and Transportation Company of Colorado was formed in 1876 and the Continental Oil and Transportation Company of California in 1877.

By 1878, not only was Blake selling products all along the coast, he was evolving plans for marketing oils, lubricants and waxes outside the United States. Continental became an international marketer, shipping products to agents in Canada, Mexico, Japan, the Hawaiian Islands and Samoa. Manufacturing facilities were established in San Francisco and Los Angeles to meet the export demand. Looking for a way to move oil to San Francisco, in 1883, Blake directed the construction of the first pipeline in California, from Pico (near Santa Clarita) to Ventura, where oil was loaded onto steamers for transport to the Golden Gate.

While these operations proceeded smoothly, the company turned its attention to the Rocky

HORSE-DRAWN TANK WAGONS

Continental relied on a network of bulk stations that were located, for the most part, within a radius of 15 miles from all points of distribution, both wholesale and retail. A great inefficiency remained, however, in that oil had to be transported and delivered to customers in barrels and the empties picked up and replaced. Mule- and horse-drawn tank wagon delivery, an innovation Continental pioneered, cut down on the awkwardness of having to wrestle heavy, bulky barrels. Tank wagons were fitted with 400-gallon tanks — "a regular horse killer," a former driver recalled. Customers then could keep their barrels and have them refilled on-site. Wagon drivers typically began work at four in the morning and made the rounds until nine in the evening. An average of 1,600 gallons of kerosene was delivered daily by each of the tank wagons in service during the 1880s.

Mountains, establishing a network of offices, warehouses and bulk plants throughout the region. A typical bulk station was a primitive affair, consisting of two or three large wooden storage tanks of 500-barrel capacity, a storage warehouse, a stable and a small office.

Oil left the stations in barrels for delivery to customers, most of them grocery stores with sawdust on the floor to absorb oily residues. At first, the stores were required to buy an entire barrel even if they only wanted a few gallons. Each barrel also needed to be delivered and picked up by the company when empty — a nagging, back-aching proposition for deliverymen. Blake solved these dilemmas by using horse-drawn tank wagons, which removed the need to wrestle heavy barrels on and off carts and also permitted drivers to pour out the exact volume of kerosene customers wanted.

Each tank wagon was fitted with a 400-gallon tank and had small wheels and heavy platform springs — "a regular horse killer," a retiree recalled. "The drivers wore long gauntlet gloves and hats big as umbrellas. They'd come galloping into town and stop suddenly so near the curb that they brushed the dust off the spokes without touching the paint!"

Many drivers began work at four o'clock in the morning to have their horses brushed and glistening before heading out. One driver kept a diary in 1888 recounting the dull round of deliveries he made to the 24 groceries in his territory, noting whimsically

U. S. Hollister, vice president of Continental, sits in a buggy (far right in the photo) in front of the company's Trinidad, Colorado, bulk plant in 1909. Hollister wrote of the bitter battle Continental fought with Standard Oil Company as it moved into Continental's Western territory. Ernest Whitworth Marland, opposite, favored Norfolk jackets, knickers and spats. This meticulous appearance belied Marland's economic condition when he traveled to make his fortune in Oklahoma oil. Virtually penniless when he arrived, Marland would become a wealthy oil baron rivaling John D. Rockefeller himself.

that it wasn't all work. "Only diversion while on these trips between towns was killing of rattlesnakes," F. D. Stafford wrote. "Carried a six foot length of chain for the purpose."

While tank drivers battled serpents, Continental faced a more formidable foe — John D. Rockefeller and his giant Standard Oil Company. During the early 1870s, most of the oil products sold in California came from Standard Oil's refineries in the East, arriving in clipper ships sailing around Cape Horn. But by 1879, overland rail rates were dropping, and Rockefeller and his companies decided it was time to market products directly in the state via a subsidiary, Consolidated Tank Line Company. In 1879, he opened an office in San Francisco.

This was just the kind of challenge Isaac Blake relished. An outspoken man with a strong competitive streak, he worked day and night, guided by a simple, yet profound, objective: to be the best in all he did. "When he directed the choir at Trinity Church in Denver," a friend recalled, "it was one of the finest singing groups in the country." His aim within the oil industry was no different.

Blake pronounced that Continental would drive Rockefeller out of the market. A fierce battle ensued. "We fought in the ring and outside among the spectators, out on the street, and along the country road," U. S. Hollister, Continental's vice president at the time, later wrote.

"It was a fight to the finish."

On a dry December morning in 1908, Ernest Whitworth Marland, a Pennsylvania lawyer, oil operator and self-taught geologist, stepped off the Santa Fe train in Oklahoma, loosened his tie and shook the wrinkles out of his belted Norfolk jacket. A few cowboys lounging at the depot looked him over and decided he was just another Eastern dude touring the West. But Marland's appearance — he also wore knickers and spats — was deceiving. At 35, this stubby, bustling man had already made and lost a million dollars in the oil and coal fields of Pennsylvania and West Virginia. He may have looked like well-fed visiting royalty, but he was flat busted broke.

Marland had an idea of how he would recoup his fortune in oil, and it involved the 101 Ranch. The 100,000-acre cattle ranch and farm, built in 1881 and leased from the Ponca Indian tribe by the Miller family, was among the most famous in America. The spirit of the old West lived on at the ranch, and its touring Miller Bros.' 101 Real Wild West Show — kicked off in 1906 and featuring Tom Mix, Bill Pickett, Hoot Gibson and other cowboy notables — competed against Buffalo Bill's Wild West extravaganzas.

Marland's half-cousin and boyhood friend Franklin Kenney, a Spanish-American War veteran and a second lieutenant in the U.S. Army, was stationed at Fort Sill, Oklahoma. There he had met and befriended the Miller brothers: Zack, Joe and

THE MILLER BROTHERS AND
THE 101 RANCH

The famed 101 Ranch was founded by Colonel George Washington Miller in Oklahoma around 1881 and hit its heyday under Miller's three sons, pictured at left. The ranch spanned some 110,000 acres at its peak and included the largest diversified farming operation in the United States. Within its confines were several churches and schools, three towns and miles of roads — all of it purchased in lots from the Ponca Indian tribe. So vast were its holdings that the ranch published its own newspaper, the *Bliss Breeze*, and printed its own money. The 101 Ranch was famed for more than just its size. The 101 Real Wild West Show left the ranch in 1906 with 126 performers, teams of horses and equipment and toured the country, showcased at both Chicago's Coliseum and New York's Madison Square Garden. In 1908, E. W. Marland surveyed the Miller's ranchland and determined that oil lay beneath. He mapped the terrain and secured a lease from the Miller brothers to drill for oil on some 10,000 acres.

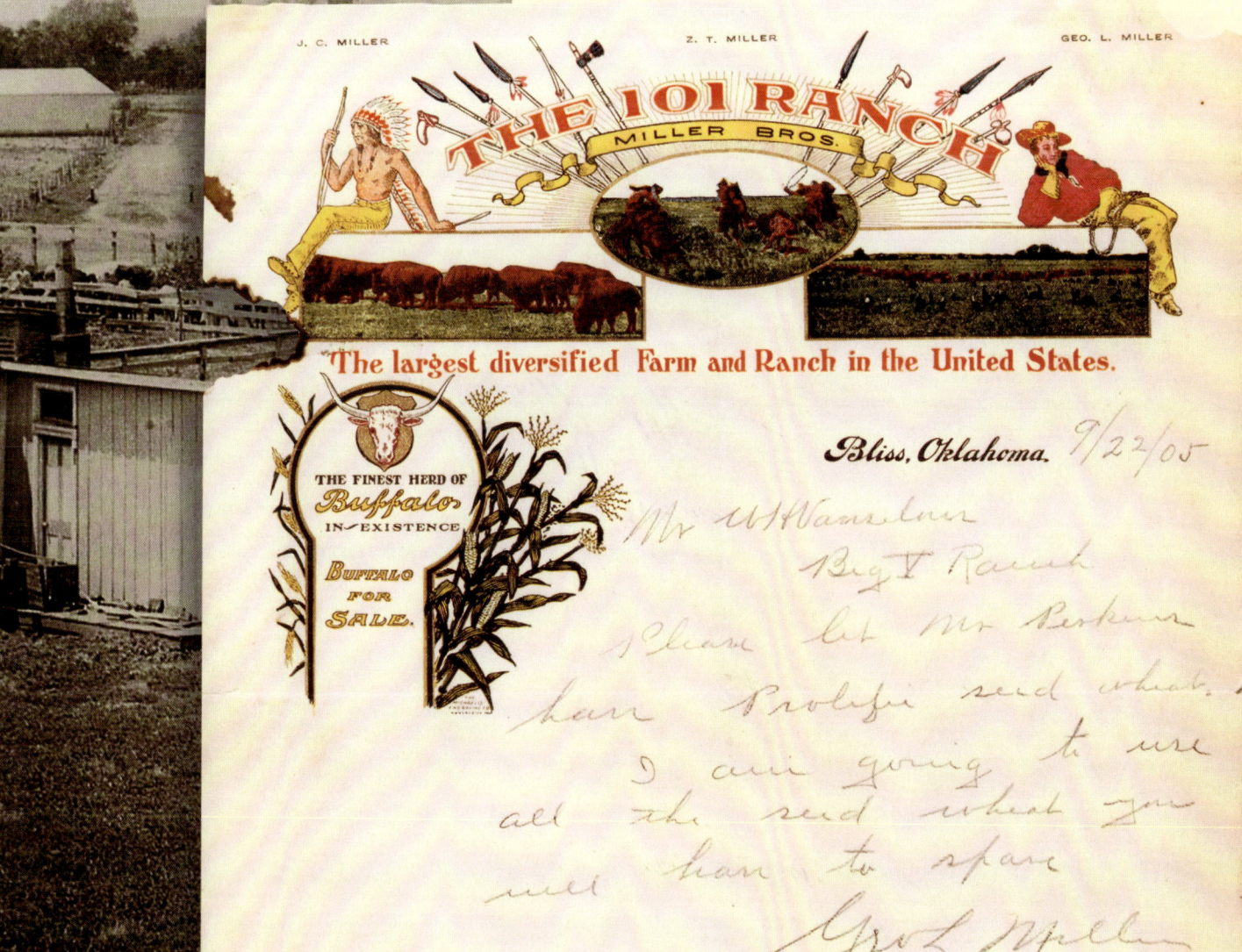

George. "The brothers believed there was oil on their land, and they persuaded Kenney to write [Marland] to come take a look," said Leland Smith, a Conoco retiree in Ponca City and an avid historian of the ranch. Eager to escape his failed business enterprises in Pittsburgh, Marland headed West.

On the day of his arrival, Marland joined Kenney for a ride out to the Millers' ranch headquarters, set in the middle of the prairie. In the years to come, Marland would tell friends that his spirits lifted as he rode across the sunlit Oklahoma fields. He had seen a prominent outcropping of rock just south of Ponca City, then a small trading town that was "no more than a wide place in the road," a former Marland Oil employee recalled. The rock formation suggested oil beneath.

Marland was a die-hard disciple of the nascent science of geology, which, he was convinced, would revolutionize the hit-or-miss nature of oil drilling. He mapped the immediate terrain and secured a lease from the Miller brothers on 10,000 acres of land to drill for oil. The agreement marked the beginning of the 101 Ranch Oil Company.

Marland also had his eye on a

The 101 Ranch show featured prominent Western stars of the period, including Bill Pickett (at left in this photo), a top "bulldogger" who was of African-American and Native American descent. Other performers included Hoot Gibson and Tom Mix. Like its rival, Buffalo Bill's Wild West Show, the 101 extravaganza included Indian "massacres" and jaw-dropping rodeo stunts, like the rope trick featured opposite. The curtain came down on the eclectic show in 1932 because of mounting debts. By that time, oil found on lands leased from the Millers and the Ponca Indians had made Marland rich. Below is an original lease from 1912.

PARALLEL PASSAGE 27

Marland's first gusher came in 1911 on Ponca land. Chief White Eagle is, in a sense, a founding father of Conoco, having leased the land to E. W. Marland. White Eagle was an old friend of the late Colonel Miller and his three sons. When the Poncas were moved to a temporary reservation and endured sickness and famine, the Miller family fed them. So when Marland asked for permission to drill along the slope of Bois d'Arc Creek, sacred land owned by the Poncas, the Miller brothers intervened on his behalf with White Eagle, who assented to the drilling.

4,800-acre parcel of land owned by the Ponca Indians. With the help of the Millers, he arranged a meeting. The negotiating proved difficult, however. "We had a lot of troubles with the Indians, [but] after a lot of squatting, smoking and palaver, I obtained the right to drill," Marland later wrote.

The first well he drilled was under the most adverse conditions imaginable. Since there were only light horses and ranch ponies in Oklahoma, lumbering teams of oxen had to be used to haul rig timbers, tools, boilers and casing from the railroad to the well location. Marland himself frequently drove the oxen. The exertions proved futile — the well was dry.

The next seven wells hit gas, however, and the 101 Ranch Oil Company immediately sought a market, building a pipeline to Tonkawa, 15 miles (24 kilometers) away. Later, it began a small gas distributing company, Kay County Gas Co. Enough income trickled in from this venture to allow Marland to continue his search for oil, although he relied on others, including another oilman, Lew Wentz, and a partner, W. H. McFadden, to bear part of the financial burden.

Marland worked furiously, often cooking for his crew or trudging alongside the teams of oxen to deliver pipe and supplies. He rode about the ranch looking for new drilling sites, his trained geologist's eyes hunting for promising rock formations. On one such trip with George Miller, he spied some Ponca Indian graves at the top of a hill near Bois d'Arc Creek. "George was showing me around the Ranch one day, and as we rode up a hill to see the Ponca Indian Cemetery, where the Indians placed their dead upon wicker platforms above ground, I noticed by the outcropping rock that the hill was a geological as well as a topographical high," Marland later recalled.

"Marland smelled oil," said Leland Smith. "But this was sacred ground to the Poncas. So he and George Miller arranged a meeting with White Eagle, chief of the Poncas. After explaining their purpose and the promise of the venture, they got permission from White Eagle, who trusted and liked the Miller brothers, to drill along the slope. He warned them, however, to stay away from the crest of the hallowed hill, where Indian bodies were buried. Marland gave his promise and never broke it."

Marland obtained a lease to drill on the land from Willie Cry, the Ponca who owned the parcel. On June 27, 1911, the well came roaring in. A major new field, right in the heart of the 101 Ranch, had been discovered. And E. W. Marland was on his way to new fortune.

"I remember White Eagle standing near the derrick of the first well when it blew in with a terrific roar," Marland later wrote. "He told me in the sign language that I was making 'bad medicine' for him, his people and myself."

The chief's eagle eyes had a bead on the future.

Formerly the domain of the upper classes, motor-touring cars had taken the American public by storm thanks to Henry Ford's revolutionary assembly-line manufacturing process. Gasoline became the stock in trade of oil companies. Marland Oil was an early believer in the power of advertising...and the Wild West image, as this 1927 ad in the *Oklahoma Farmer-Stockman* illustrates.

Isaac Blake loved to bedevil his competitors, gleefully pinpointing their weak spots, particularly those of his nemesis, Standard Oil. When he discovered, for example, that Standard's Consolidated Tank Line Company was having problems with its lubrication oils, he hired away the company's lubrication specialist. The new recruit advertised the weaknesses of his former employer's products, quickly transferring a good many Standard customers to Continental.

Such "swoop and shoot" tactics had only modest success, however. Though Blake constructed additional bulk stations along railroad tracks to improve Continental's distribution, John D. Rockefeller's clandestine agreements with the railroads to obtain cheap shipping rates proved too much for the smaller company. Its future appeared dim. So, in deference to the old maxim, "If you can't beat 'em, join 'em," Continental hopped on the Standard bandwagon.

On January 1, 1885, the Continental Oil Company became part of the Standard Oil Trust with a paid-in capital stock of $300,000 — a decided improvement on the $25,000 originally invested by Blake and his associates 10 years earlier. The new company formally took over all holdings of the Continental Oil and Transportation Company, with the state of Colorado and the territories of Utah, Wyoming, Montana and New Mexico named as its region and Isaac Blake its president.

To the left are tank wagons of Continental kerosene leaving the company's original Denver warehouse in 1884. Denver became a Continental marketing center early in the company's history. Shortly after the company's founding, Isaac Blake and his associates went to the Mile High City to consider potential markets in Colorado and lay plans for opening offices there. Above is an advertisement from the same period showing Continental's position as a top distributor in the West. Continental proved no match for John D. Rockefeller's Standard Oil, however, which had obtained exclusive shipping discounts for its Eastern oil. It would become part of the Standard Oil Trust January 1, 1885.

PARALLEL PASSAGE 31

An early advertisement pictures a skyscape of oil derricks. Although Continental did not yet produce oil and was involved only in its distribution, marketing and transportation, the company's merger into Standard Oil allowed it to pour more dollars into advertising. Salesmen even carried thermometers with the legend "Continental."

Shortly after the merger, the company expanded into Arizona, Idaho and Nevada Territories, the Pacific Coast, Alaska and the northwest part of Canada. It still focused on marketing and distributing products bought from other companies but with a much larger fleet of vehicles and tank cars, thanks to Standard — indeed, the largest assembled within one organization at the time.

But Blake was restless. Now that others were calling the shots, he began to focus his energies and money elsewhere — on the railroad business, which he saw as the arena of action for the 1890s. His timing was terrible: The financial panic of 1892 decimated monetary values, and he lost more than $1 million of his personal wealth. In August 1893, burdened with debt, he resigned as president of Continental.

Through his remaining days, Blake pursued the same exhilarating trinity of risk, reward and ruin. "His intense mental activity led him into other lines which he understood little about, and that with his natural optimism ultimately cost him his entire fortune," Franklin Morgan, the company's secretary in 1889, later wrote.

The depression that followed the Panic of 1892 took a toll on Continental's employees. The company helped out where it could, adding a commissary in its main Denver offices — "a small plant located at 11th Street and the railroad tracks," a memorandum stated — where employees could buy potatoes, flour and other staples wholesale. One retiree recalled paying a nickel for 10 pounds of potatoes in 1898, a pretty good bargain, given his salary was $1.50 a day, paid in silver.

Henry Morgan Tilford, an oil dealer from Kentucky who managed one of Standard's New York departments, was elected president of Continental upon Blake's resignation. Portly and austere, Tilford spent much of his time in New York, where Standard made its headquarters, traveling to Continental's marketing region only occasionally.

B. G. Wilson, Continental's New Mexico manager at the time, recalled one of Tilford's sporadic visits to Denver. "I took him to Chase's Palace Theater," Wilson later wrote. "A man in the seat right smack next to Tilford began to needle the singer, proclaiming loudly that he sang like a bull-calf. The singer, professing great indignation, rushed back into the wings, snatched up a large six-shooter and began shooting blank cartridges at his confederate in the audience. Tilford turned white and bolted for the door." The sophisticated New Yorker didn't realize it was all part of the show. No wonder he stayed in the East.

To lure greater market share, Continental turned to advertising its products. Salesmen carried thermometers, pencils and memorandum books with "Continental" stamped on them. A newspaper advertisement promised that Continental Safety Oil (plain kerosene sold in bulk) was "GUARANTEED equal to best cased oil [kerosene] in the market." Another advertisement, published in 1900, reads: "A good lamp is a blessing! A good oil heater is a great convenience, and a successful oil cooking stove is a luxury, and a saver of time and money, Continental Oil Company — Specialty Store, Corner 15th and Tremont Streets, Denver."

As the advertisement indicates, Continental was a rather eclectic distributor at the century's turn. Besides lamps and stoves, the company sold candlesticks, birthday candles and even paraffin chewing wax before the invention of bubble gum. Prices on some lamps exceeded $100, and many of the better grades of candlesticks and lamps had silk shades and sparkling glass-beaded fringes.

The company also emphasized service: In Denver in the late 1890s, Continental employees rode bicycles to customers' homes to clean kerosene stoves and trim lamp wicks!

During the Spanish-American War, some employees fought with Teddy Roosevelt's "Rough Riders" in Cuba, while the company back home supplied harness oil, axle grease and candles to forts in Denver and Laramie. Candles for the mining industry were another growing line, as "no other means of lighting the mines had been developed until some years later," noted Emerson G. Smith, the company's Denver-based bookkeeper, in 1907. Mining companies purchased candles "by the carload," Smith added.

These were prosperous days for some — a chicken in every pot, sleek horses in every stable and, here and there, a shiny new gasoline buggy. And there were world events of importance: The Paris Exposition opened in April 1900. The Boxer insurrection bathed China in blood and intrigue. And on Wall Street, Hill and Harriman, the kings of the railroad, battled for control of the Northern Pacific, cornered the stock and brought on the financial panic of 1901. There went the chicken.

Continental was still a frontier company, however, far from the refinements of the East. A division manager in Denver at the end of the nineteenth century recalled having a nice chat and a smoke with two strangers, one of whom turned out to be the

Pictured is the Denver office, which became Continental's headquarters in 1884, the same year in which Blake and his partners accepted an offer from Standard Oil to become an affiliate. Blake resigned from the presidency in 1893 and was replaced by a Standard Oil man, Henry Morgan Tilford, below.

Continental Oil's 1908 baseball team earned the title of Rocky Mountain champions. The player in the front row second from left is W. H. "Billy" Potter, who came to Continental as a 13-year-old office boy in 1898. The engraved silver matchbox was a gift to Continental customers in 1903. This old-fashioned "lighter" held wooden matches; the bottom was used for striking.

infamous outlaw Butch Cassidy, wanted "dead or alive" at the time for a nifty $5,000 reward.

As the new century progressed, some Eastern pastimes did find their way west. At least nine Continental employees formed a baseball team, their uniforms emblazoned with the letter "C." Others joined the Continental Oil Wheel Club, which competed in road races up to 100 miles (161 kilometers) — the so-called "century runs." Onlookers at both events often sported "cast-iron" derbies, high-top button shoes and celluloid collars — the modish Eastern fashions.

Back at the company's main offices at the McPhee Building in Denver, Continental had but one telephone, located in a small booth in the hall near the office boy's desk. "It was necessary for me to climb on a tall stool to reach the telephone when calls came in," recalled Billy Potter, an office boy in 1898.

By the time 1906 rolled around, Continental, with roughly 60 distribution points in the West, had captured 98.9 percent of the Western kerosene market. The following year, 38-year-old Edward T. Wilson became president and chief executive officer, a post he would hold for 17 years. A self-made man, Wilson started in his teens with Standard Oil and moved rapidly up the ladder. Unlike Tilford, he made his offices not in New

The Continental Oil Wheel Club posed for a shot just before the first annual road race from Denver to Littleton, Colorado, June 20, 1895. The club competed in road races up to 100 miles, called "century runs."

York City, but in Denver's McPhee Building, which he renamed the Continental Oil Building.

Business was booming, but not from kerosene, lamps or cooking stoves. Stanley Steamers, Maxwells, Duryeas and other motorcars were plying the country's roads in larger numbers — from 300 in 1895 to 460,000 in 1910 — pumping up demand for gasoline.

At the time, most garages handled the fuel in small portable tanks, which were wheeled to an automobile and pumped directly into the tank. Not for long: In 1909, Continental positioned two large hot-water boilers on a warehouse platform to deliver gasoline by gravity feed through a tube to the tanks of waiting motorcars. "Cars would be lined up and wait for 30 to 45 minutes to be served," Emerson Smith recalled. "I believe this could be considered the first filling station in the West, although Standard had some filling stations on the [East] Coast."

The company began investing in motor trucks for transporting fuel, each truck equipped with three tanks to deliver different types of fuel. Unfortunately, there were no delivery hoses on the vintage 1910 trucks. "We used to have to bucket off the gasoline, five gallons at a time," a retiree recalled. Moreover, the trucks kept breaking down. If the tanks weren't evenly filled or emptied, chances were the front end or rear end would lift off the ground!

The year 1913 was a pivotal one for the company. On March 31, following a Supreme Court ruling two years earlier requiring Standard Oil to dispose of holdings in subsidiary companies, the Continental Oil Company of Colorado was incorporated with a capital stock of $3 million. Continental was an independent once again.

As Edward T. Wilson looked to the future that year, he realized it spoke one word: gasoline. Cars were everywhere, it seemed, more than 1.7 million of them thirsty for the old stove-cleaning product benzine. Wilson directed the establishment of Continental's first service stations — tiny square steel or brick buildings featuring fuel pumps in which the product was visible. The gasoline flowed into a large glass cylinder so customers could actually see it being pumped — a novelty. By the early 1920s, signs at the company's 250 stations featured the company's trademark, a Continental soldier, encircled by two words: Conoco Gasoline. The Conoco name became a familiar one to consumers, though the Continental Oil Co. would not change its official corporate designation to Conoco until 1979.

But gasoline retailing wasn't the only indication that Conoco was a new company. Within a few years, the pioneer marketer would become a great producer and refiner of oil as well, competing against the still-powerful Standard Oil companies and other newcomers to the field, one of them an upstart called Marland Oil.

GREEN, MERCANTILE CO. BUFFALO CREEK COLORADO 80425

Motorized tank wagons, opposite, fueled by the very product they carried, took the place of the old horse-pulled wagons. Continental sold gasoline under the company's recognized brand name and later corporate name: Conoco. To the left is Buffalo Creek, Colorado, and the J. W. Green Mercantile Co., site of what many believe is the oldest Conoco service station, pictured sometime after the merger between Continental and Marland. Three generations of Greens lived and prospered here, selling Conoco products since 1884.

PARALLEL PASSAGE 37

Ponca City, Oklahoma, headquarters of Marland Oil, looked like a bustling Western town in 1910, when this photo was taken. Named for the Ponca Indian tribe, the city grew up with the fortunes of the Marland Oil Company and remains an important part of Conoco today. Over the years, E. W. Marland's benevolence funded the building of schools, a hospital, a stadium at the University of Oklahoma and a home for destitute children. To teach the locals polo, his lifelong passion, Marland even brought over a coach from England.

The thrill of that first oil discovery by E. W. Marland in 1911 was accentuated by the fact that every dollar he had in the world, every bit of credit he could muster, was in that hole in the ground. "It just had to strike, or I had to go work for someone else," Marland later wrote. "It meant my entire future."

The strike was the first real evidence that the mid-continent was a huge area of potential oil reservoirs — as geologists had speculated for years. Marland knew he had had the great fortune to be in the right place at the right time. He soon dreamed up plans for his fledgling 101 Ranch Oil Company — tank farms needed to be constructed, of course, as well as pipelines, office buildings, a refinery and houses for employees. And he had a town to build. Ponca City, Oklahoma, that "wide spot on the road," soon became a bustling oasis amid the wheat fields of Oklahoma.

As Marland expected, that first gusher on Willie Cry's land brought scores of oil entrepreneurs to the region. But he believed he had a leg up on the competition — the application of the science of geology to oil finding. Other oil producers scoffed. "[They] put about as much confidence in witch hazel or a peach fork as they did in the geological facts of the earth's history," John Joseph Matthews writes in his book on Marland, *Life and Death of An Oilman*.

Marland's thin geological training in the East directed him to search for anticlines, folds in the earth in which strata incline downward on all sides. Underneath such folds, the current geological studies indicated — and Marland believed — was oil or natural gas. In 1912, he met Professor Irving Perrine, a geologist from the University of Oklahoma, who refined his knowledge of geology and its relationship to oil reserves.

Perrine and several of his colleagues and students later became Marland's de facto geological department, coming up on weekends and holidays to Ponca City to discuss anticlinal theories, map rock formations and target drilling sites. "My geological staff managed to keep a little ahead of the industry by improved scientific methods which we guarded as carefully as we could from the scouting departments of the major companies," Marland wrote.

By 1916, the 101 Ranch Oil Company was producing from three Oklahoma oil fields: Ponca City, Newkirk and Blackwell. The Garber field was added the following year and the Billings field a little later. Much of the crude was promptly sold at the wellhead to large buyers like the former Standard Oil companies. But Marland had bigger plans. In 1917, he reorganized the company into the Marland Refining Company and the following year, built a small refinery at Ponca City to process the crude oil the company produced. John D. Rockefeller lost a major supplier.

With production and refining facilities now

established, the company entered the retail market in the states of Oklahoma, Kansas, Arkansas, Missouri, Nebraska and Iowa. Demand for the company's products, primarily kerosene, gasoline and lubricants, jumped during World War I and skyrocketed after the war as Americans took to the automobile.

The money generated was invested in new land leases. In November 1918, after the western Osage reservation was opened for leasing, Marland's Kay County Gas Company leased 192,048 acres from the Osage tribe. He had outbid some of the biggest oil companies in the world, including the Standard Oil companies and Royal Dutch Shell.

His plan was to lease as much land as possible, the gambler's strategy of playing the odds. "For E. W. Marland, 'speculation' was not a pejorative word," one retiree recalled. Overseas, it was the same game. In 1921, he met with Mexico's President Carranza to arrange blanket leases along the coast of Sonora, Sinaloa and Lower California. Millions of acres were leased in all. Later, a Marland employee was sent to Mexico to make a payment with $10,000 in gold in a suitcase tied to his wrist.

In the United States, however, the U.S. Department of the Interior's regulations limiting oil companies to a maximum acreage at each land-lease auction blocked such gargantuan concessions. Marland neatly sidestepped the rules by establishing various paper corporations, most of them named

By the late 1910s, gasoline already was the top-selling product of crude petroleum — thanks to the proliferation in the number of automobiles owned by Americans. Motor-touring was still an adventurous (and dusty) affair, however, and early gas stations were primitive, as the photo to the left clearly attests. Here a Model T Ford fuels up with Marland gasoline at a log cabin service station in Lanagan, Missouri, in 1919. Above, Marland's 1917 mobile "gas-testing" machine helped the company perfect its chief product.

PARALLEL PASSAGE   41

Petroleum, a booming business in America, was processed into some 300 different products, including benzine, naphtha and asphalt. Marland Oil was a beneficiary of the industry's might, its prospecting teams bringing in one giant oil field after another. By 1921, the company's oil holdings were valued at more than $100 million.

after his friends. One, the Kenney-Cleary Company, honored his half-cousin Franklin Kenney. Each "company" had offices in the Marland office building, yet they bid separately at auctions.

By 1922, there were so many subsidiary and arm's-length branches of Marland's growing empire, he was compelled to form the Marland Oil Company as a holding corporation. The red triangle he adopted as the company's emblem in 1919 now featured the legend "Marland Oils" in the center. In 1920, Marland hung a sign with the logo at his first service station, in Pawhuska, Oklahoma.

Postwar Americans invested their energies in frivolity — the delirious decade of flagpole sitting, hip-pocket flasks, Ruthian exploits and bathtub gin known as the Roaring Twenties. Although this was also the era of Prohibition, one would not know it in Ponca City. Marland "had the best bourbons and ryes" and got his Scotch and champagne "right off the boat," the biographer Matthews writes.

Petroleum was America's greatest asset during the boom period, accounting for $4 million in exports. More than 300 different products were processed from crude oil, including benzine, naphtha, asphalt and even soap. In 1919, isopropyl alcohol was synthesized from refinery gases, marking the beginnings of the petrochemical age, though it would be decades until the full economic promise of petrochemicals was realized.

Money flowed through Ponca City like the pools of oil beneath. But greater fortune lay ahead. On May 14, 1920, at 2,965 feet (904 meters), Marland's drilling crew struck oil about 20 miles (32 kilometers) east of the Ponca City refinery. They had discovered the giant Burbank field. "Soon, where there had been no sound except the doleful whistle of the upland plover," Matthews writes, "there developed the metallic rhythms of drilling, the coughing of pumps and the explosive laboring of trucks stuck in the mud."

Marland and his geologists had no way of anticipating the value of the discovery. Lease sales for the Burbank field reached unprecedented heights, with some companies bidding almost

A smiling Marland (seated second from left) gathers with a group in front of the Willie Cry well, the first Marland well to hit oil, above. This and the other early producers of the 1910s, such as the modest derricks in South Ponca field, opposite, pose a stark contrast to the view of Three Sands field near Tonkawa, Oklahoma, in the mid-1920s, on pages 44 and 45 following.

Marland gas stations featured clean facilities that blended into the local environment. The logo was reportedly derived from a Scottish emblem of hospitality. The three corners of the triangle symbolize quality, service and courtesy. Marland also had a reputation as a generous employer, instituting benefits, such as insurance, that were ahead of their time.

46 125 YEARS OF ENERGY

$2 million for one 160-acre tract. Then, in June 1921, Marland's crews brought in another behemoth, the Tonkawa field, 15 miles (24 kilometers) southwest of Ponca City. The back-to-back finds spouted more than 100,000 barrels of oil daily.

With his oil holdings in 1921 valued at more than $100 million, Marland was dubbed "the millionaire producer of Ponca City" by a United Press International correspondent. He spent his money generously on his adopted city, building schools, a hospital, city parks and a home for destitute children. He bought a bank in Ponca City so he could lend employees money at 6 percent interest when the standard rates in local banks were between 8 and 10 percent. And he paid his employees "not only a living wage, but a savings wage," he once said. "I spent money on my people and my town, and they both flourished." By 1922,

PARALLEL PASSAGE 47

As motorists lined up to fill their tanks at one of the many Marland stations dotting Oklahoma in the early 1920s, the company gave them detailed maps of the region that included other Marland stations. This Marland station in Oklahoma City set a world's record in the early 1920s by pumping 20,582 gallons of gasoline in one day. The U.S. Geological Survey glumly predicted in July 1920 that the country's oil supply would be exhausted in 17 years if taken from the ground at the 1919 rate of 400 million barrels per year. By 1927, U.S. production was less than 40 barrels for each automobile registered, down from 8,000 barrels in 1900.

Marland controlled a great deal of the world's known oil, and more than one-third of Ponca City's population was employed by the Marland Oil Company.

And, boy, did he live lavishly! He owned a yacht, a private railroad car and one beautiful estate after another. A devotee of both polo and English fox hunting (after all, he already had the clothes), he brought both urbane pleasures to Ponca City, causing the natives to goggle. But he worked as hard as he played, and from 1921 to 1927, he built the Marland Oil Company into a substantial integrated oil company. He sent teams of geologists and drilling rigs to California, Colorado, Louisiana and Mexico, enlarged his modest refinery and constructed a chain of service stations that resembled tiny English cottages — no grease-monkey pits for lifelong Anglophile E. W. Marland!

And though it seemed the gambler's luck would never end, he lost his way among his riches.

SERVICE STATIONS
OF THE 1920s

Polo-playing Anglophile E. W. Marland directed in the mid-1920s that company gas stations adopt the look of English cottages. Hundreds of the appealing stations were built during the 1920s and 1930s, although only a few remain today. The tradition Marland initiated was upheld by later leaders. A 1934 advisory to station managers recommended that flowers be planted on the premises to "further beautify" the area. A couple years hence, a few stations included early car-washing facilities. In 1926, Marland achieved a personal ambition of sorts, acquiring the Sealand Petroleum Company to distribute Marland Oil products, like the one at left, in the United Kingdom.

Free at last from John D. Rockefeller's grip, Conoco set its sights on entering the oil production and refining fields. The rationale was simple: By refining its own crude oil, the company would cut out the profits of refinery middlemen and, thus, be able to price its products more competitively. The hunt began in earnest for a refinery to purchase.

Edward T. Wilson, Conoco's president and CEO, found what he was after in the company's file drawers. Back in 1888, Continental had acquired a minority interest in the United Oil Company, a small producer and refiner in Florence, Colorado, site of the West's first oil field and refineries as early as the 1860s. Continental had agreed at the time to distribute United Oil's light products. This serendipitous decision bore fruit in 1916, when Conoco acquired United Oil in full — lock, stock, barrels and, of course, its refinery.

Conoco now needed to produce some crude oil to refine. In late 1916, the company entered the Wyoming oil fields as a producer through a subsidiary, Continental Oil Producing Company. Although both its refining and producing enterprises were modest at best, Conoco was now a fully integrated oil company capable of handling the entire petroleum cycle.

That same year, a new company was formed that would provide the vital production capacity Conoco was looking for, the Elk Basin Petroleum Co., incorporated in Maine on December 11. The company initiated a string of acquisitions through the 1910s and 1920s that provided tremendous growth during relatively lean years in the industry. It acquired a constellation of companies, including Keoughan-Hurst Drilling Co., Mutual Oil, Frantz Corp., Western Oil Fields Corp., Boston-Wyoming Oil Co., Hamilton Oil, Merritt Oil, United Oil and Sapulpa Refining Co. By 1924, Elk Basin, renamed the Mutual Oil Co., was eyeing Conoco.

It was Conoco's marketing strength that made it a target of this campaign. In 1917, for instance, Conoco had undertaken a historic venture with Yellowstone National Park. The park was in the process of replacing its horse-drawn carriages with a new fleet of motor-touring cars. Conoco became the exclusive supplier of gasoline and motor oil to the park, an agreement that continues uninterrupted to this day.

Although Conoco had entered production and refining following the buyout of United Oil, "the company's crude supply and refining facilities were insufficient for its marketing requirements," explained S. H. Keoughan, the brains behind Mutual's successful acquisition strategy. Mutual Oil looked like a partner that could fill the gaps. By 1924, it was producing 16,000 barrels of crude oil a day and refining more than 11,000 barrels a day at its Glenrock, Wyoming, facility. Moreover, it was marketing products in Kansas, Nebraska, North and South Dakota, and the Rocky Mountain region.

That spring, the Mutual Oil Company acquired a majority interest in the Continental Oil Company (Conoco) and formed a new company, the

In 1917, Conoco undertook a historic venture with Yellowstone National Park to be the exclusive supplier of gasoline and motor oil to the park, an agreement that continues today. Visitors had a reliable supply of fuel as they toured one of the nation's top destinations, opposite, as did the park's own fleet of motor-touring cars, left.

On March 31, 1913, the Continental Oil Company of Colorado was incorporated with a capital stock of $3 million and listed on the New York Stock Exchange, following an antitrust ruling two years earlier by the Supreme Court against Standard Oil. Pictured is a stock certificate of the once-again independent company. Conoco's Denver headquarters building, completed in 1926, was crested by one of the five largest and heaviest signs in the United States, featuring the company's trademark soldier, opposite. After the merger of Conoco and Marland Oil in 1929, the sign was replaced by the company's new logo, borrowing the triangle from the Marland logo and incorporating the legend "Conoco." At the dedication ceremonies of the new sign — a full eight stories tall — dancing girls climbed it for a publicity picture.

54   125 YEARS OF ENERGY

Continental Oil Company (Maine) to take over its assets. The new Conoco, managed by Mutual executives, combined the valuable production capacity of Mutual and the marketing strength of the old Conoco. "The merger put us in the production and refining end of the business in a big way for the first time," recalled Ed Snell, a statistics clerk in Conoco's Denver office in the 1920s.

Negotiating the merger turned out to be Wilson's final and finest act in office. On March 20, 1924, just 19 days after the stock swap between the companies, Wilson retired. C. E. Strong, who began his career with Conoco as a general accountant in the Denver office, was elected to succeed him.

In 1926, the company completed a new headquarters building in Denver, at 17th and Glenarm Street. The tall, white building — considered an architectural marvel at the time and a beacon for early aviators — was crested by one of the five largest and heaviest signs in America, "so ponderous its supports were sunk all the way to bed rock, more than ten floors below," company historians reported in 1975. The sign featured the company's then-familiar trademark: the Continental soldier.

Jen Sellers, who joined the company's personnel department in 1926, saw the building as unique for the time. "All the hallway floors were marble, and I remember how the office boys roller-skated from one office to the next, delivering mail," recalled Sellers in a 1970s interview. The building

also featured a large auditorium on the ground floor, where employees staged dances and plays and watched silent movies, "until the talkies arrived later in the decade," Sellers added.

Keoughan was elected Conoco's president and chief executive officer in 1927. As the 1920s progressed, Conoco bulked up its refining and production capacity and expanded the number of its marketing outlets, acquiring the Texhoma Oil and Refining Co. The company was actively drilling in many locations in the West and operated eight refineries in six states. It owned more than 1,700 producing wells and 530 miles (853 kilometers) of pipeline. By the end of the decade, Conoco filling stations numbered more than 1,000 in 15 states.

But Conoco's greater size did not necessarily imply greater profitability. To compete effectively in the rapidly consolidating industry, Conoco would require more capital, greater assets and, most importantly, added strength in exploration and production.

Conoco needed a more powerful partner.

In the early days of the company's international and domestic shipping enterprises, the custom was to name vessels after prominent Conoco executives. Pictured below left is the *E. J. Nicklos*, an early tanker. Conoco's enviable distribution network drew the attention of the Mutual Oil Co., a producing and refining company that grew by acquisition through the early 1920s. In 1924, Mutual and Conoco joined forces under the Conoco name, but the company would need to ally itself with even greater producing power to compete in an increasingly competitive market. The Marland Oil Company had that power and immense refining capabilites as well. Marland's Fleming Stills, located in Ponca City, right, was the largest battery in the country.

E. W. Marland's reliance on geological mapping as a means to discover oil served him well, as the company racked up one major discovery after another with the help of technical staff. In "the 1910's," Marland established the first research division in the U.S. oil industry. The division developed the first core drill to evaluate oil deposits and brought over from Germany refraction seismography techniques (and scientists) to find oil. Pictured above is the Marland control lab in Ponca City in 1926. Marland also poured money into the Ponca City refinery and tank farm, built in 1918, opposite

Marland's surefire touch — his uncanny ability to stick a spigot in the earth and pump up oil — stupefied other oilmen, who considered him a maverick. Although his belief in geology as a tool to discover oil was more alchemy than pure science, Marland's head was in the right place. Technology would change the oil business, and he was among the few to realize its significance.

He established the industry's first research division in the late 1910s, staffing it with more than a dozen scientists. The division paved the way for a series of technological advancements, including the use of the core drill to evaluate oil deposits. Marland had seen core drills used back East to determine the thickness of coal beds and felt they had application for the oil business as well.

The company also was the first to realize the oil-finding potential of refraction seismography. The early days of this technology, invented in Germany, involved the detonation of a series of dynamite explosions in areas deemed promising oil deposit sites. Waves emanating from the center of the explosion through the strata were then measured and charted to determine the shape of the rock beneath. In 1921, Marland assembled the first refraction seismography crew in the U.S., importing the German engineers who discovered the technology and teaming them with his geological staff.

"I must have ruined half the settings of eggs in

the West with my miniature earthquakes," Marland later wrote.

The new methods paid off handsomely, and before long, the company was awash in crude oil. Its refinery was bursting at the seams, however, and Marland was forced to deliver 30,000 barrels a day to the refineries of Standard Oil, New Jersey, a proposition that tormented him. To generate the dollars needed to expand his refinery's capacity, Marland formed a joint venture with Royal Dutch Shell in 1921. Called Comar Oil Company, the new enterprise gave Marland access to Shell's exploration methods and turned a nice profit, as well — more than $25 million over the next five years.

He also searched the Gulf Coast for a suitable outlet from which to ship his refined products to international markets, settling on some land in Texas City, Texas, in 1922. The port city had a direct feed from Ponca City via the Santa Fe Railroad.

Back home, Marland finally enlarged the Ponca City refinery's capacity in 1922, building it from 8,000 barrels a day to 14,000 barrels. In reaction to low oil prices (10 cents a barrel), he also ordered 24 new 80,000-barrel-capacity storage tanks, thereby increasing overall storage capacity to nearly 2 million barrels. Both ventures required a considerable and immediate outlay of money, at least $1.6 million in refinery improvements alone. "I borrowed from all the banks with which I had business connections in Oklahoma, Kansas City, St. Louis and Chicago, and

In the guise of rescuing Marland Oil from debt, financier J. P. Morgan Jr. bought $12 million of the company's stock in return for representation on its board of directors. Ultimately, Morgan's powerful will exerted itself on the board, and as Marland Oil's losses rose in the late 1920s, Marland himself was forced from power. Here a visibly strained Marland (center) is seen toward the end of his presidency. On November 1, 1928, he resigned.

one bank in New York City," Marland wrote.

He established a $5 million line of credit with the banks — not a great amount considering that the company's monthly gross income was greater. Then, in 1923, Marland received the phone call that would change his life. The great financier and plutocrat himself, J. P. Morgan Jr., wanted to meet him. "I had a very pleasant meeting with Mr. Morgan and his associates," Marland later wrote.

"They asked me if they could be of help to me by establishing a line of credit for me in their bank, [and] suggested that I could close my lines of credit in Chicago, St. Louis and Kansas City. This suggestion, besides being very flattering, was very agreeable to me."

Marland was anxious to get into the producing business in Texas, New Mexico and California. He also wanted to further expand the Ponca City refinery. Altogether, he figured he needed at least $12 million. Rather than lend such a large amount, which even Marland agreed would be difficult to repay in short order, Morgan and his associates dealt another hand: They would buy $12 million of stock in Marland Oil, in return for representation on its board of directors.

Morgan assured Marland he had very little interest in the oil business per se, but just "wanted to learn something about it," Marland recalled. Shortly thereafter, three Morgan associates were elected to join the company's other 12 board members. Chief White Eagle's warning was far from Marland's mind.

At first, the decision seemed prudent. The cash infusion helped fund experimental exploration technologies, expand pipeline facilities, enlarge the fleet of tank cars to 3,100 and build more than 500 "modern" service stations, each with "toiletries that aid milady's beauty," company literature proclaimed. It also gave Marland the financial clout to lease even more land: In 1926, total leased acreage shot up 60 percent over the previous year.

Marland brimmed with confidence. Everything he touched, it seemed, turned to black gold. In 1925, Marland Oil discovered the Howard-

Ernest Witworth Marland was one of the fathers of modern oil exploration. He built an oil empire on new scientific methods and old-fashioned determination. In an ironic turn of events not unusual in the oil industry, the company Marland founded was wrested from his grip while he was still full of ambition. He turned to politics, becoming a U.S. congressman and the 10th governor of Oklahoma, touring the state in a custom-built limousine fit for royalty. In 1939, he returned to his adopted hometown of Ponca City, where he remained until his death in 1941.

Glasscock and Big Lake fields in western Texas, the location of the oil- and gas-rich Permian Basin. A year later, the company brought in its version of Old Faithful, a 1,240-barrel-a-day well, at a depth of 4,400 feet (1,341 meters), in Seal Beach, California. Headlines trumpeted the discovery of this new field in the Los Angeles Basin. That same year, Marland negotiated a deal — over tea and crumpets, no less! — with the historic Hudson's Bay Company to explore for oil in Canada. He also acquired a British branch for Marland Oil, the Sealand Petroleum Company Ltd., to distribute gasoline in the United Kingdom under the name "Dominion Motor Spirit."

"He was walking the earth like a king in the company of bankers and brokers, wearing his crown with the ease of hereditary royalty, and viewing the nation's prosperity with personal pride," writes Ruth Sheldon Knowles in her book, *The Greatest Gamblers*. At the end of 1926, however, Marland Oil's books showed total current liabilities exceeded $8 million, nearly $5 million more than in 1925.

This sour note barely gave Marland pause. "He was a wildcatter with a wildcatter's dream," Matthews writes. Money was poured into monuments: life-size statues of himself and Lydie — once his adopted daughter, now his trophy wife; a $2.5 million palatial estate on 320 acres, replete with satin-gaitered servants, a swimming pool and formal gardens planted with rare trees and shrubs; and,

PARALLEL PASSAGE 61

LYDIE MARLAND
AND THE MARLAND ESTATE

In 1916, E. W. Marland and his wife, Mary Virginia, adopted her niece and nephew, Lydie and George Roberts. Lydie lived in the mansion with the family and, after the death of her aunt (Marland's wife), became E. W.'s constant companion. In 1928, after having her adoption annulled, the two were married. He was 54, and she was 28. The tabloids leered. The couple moved into Marland's palatial 55-room, 12-bath mansion in Ponca City, filled to its gold leaf ceiling with fine art, antiques and crystal chandeliers. Outside, one found a series of man-made and natural lakes, Japanese sunken gardens, waterfalls and some three miles of roses along a high stone wall — all cared for by 85 gardeners.

After resigning from the oil business, the paternalistic Marland started his second career in public service. He poured money into his community, organizing and financing a sculpting competition in 1930 to commemorate the women who had made the arduous trek west to build new lives and homes. Bryant Baker, pictured at right, won the competition, and his statue, *Pioneer Woman*, was unveiled on land adjacent to the Marland estate. Marland bequeathed both the statue and the surrounding land to the state of Oklahoma, and the national treasure remains on view in Ponca City.

When E. W. was elected Oklahoma's governor, the local press referred to his First Lady as a "princess," but her fortunes would take a pathetic turn. Upon Marland's death in 1941, Lydie withdrew from society. In 1953, she packed her few remaining possessions and left Ponca City without comment.

More than 20 years later, Lydie returned to help save the mansion that had been her home. Her letter to a local newspaper in 1975 urging Ponca City to buy the estate motivated an overwhelming vote by its citizens in favor of the purchase. Conoco paid roughly half of the $1.4 million price tag for E. W.'s dream palace, which today is a museum.

Lydie Marland died in 1987 and was buried next to E. W.

The fifth floor of the old Marland Oil building in Ponca City, Oklahoma, still bears the traces of E. W. Marland, who presided there in the 1920s. John Duncan Forsythe redesigned the executive offices in 1928, including Marland's private office. Marland himself had directed an ironworks in London to craft the silver chandelier, sconces and fireplace andirons in his office. The fireplace was black Italianate marble. Opposite, leaders of the oil industry, including Marland in the white suit, gathered in the newly designed boardroom for a meeting to discuss cleaning up the oil patch. It would not be long after that the power struggle for Marland Oil would end with Marland's resignation.

64   125 YEARS OF ENERGY

later, an indisputable work of art, Bryant Baker's magnificent bronze sculpture *Pioneer Woman*, a national treasure still on view in Ponca City.

But J. P. Morgan would take a hammer to Marland's kingdom. One of the first decisions of the new board of directors was to form an internal executive committee that would have the powers of the board and relieve the other directors of the necessity of coming to New York to hold directors' meetings. The committee consisted of three Morgan representatives and three Marland representatives. Morgan was stacking the deck.

Marland gradually lost his grip on the company's direction. Several times he lobbied the committee for money to fund the building of new pipelines — each time to no avail. "Every plan of major importance I suggested for the development of the company was vetoed," Marland wrote. He was exasperated.

The company, meanwhile, was losing money, recording losses in both 1927 and 1928. Marland sensed a growing lack of confidence in his management of the company and approached a close colleague on the committee with his concerns. "He said I took too much 'human' interest in my employees, and that Morgan & Company felt that I needed under me a President of the Company…who would be hard-boiled and two-fisted," Marland wrote.

In May 1928, the committee made its feelings formally known, requesting that Marland relinquish

the president's title in return for board chairmanship. He had no choice but to agree. The committee asked him for help in finding a replacement, even though they secretly had chosen a successor, Dan Moran, then a vice president of the Texas Company.

The chairman's position would be ornamental at best, with a salary that was, in effect, a pension. "The bankers made it clear Marland would not be permitted any voice in company affairs," Knowles writes. Although the sign on the door said "Marland Oils," the company's founder and guiding spirit had become a cipher.

It got worse. The committee told Marland later that Moran could not operate the company efficiently with Marland remaining as chairman or even living in Ponca City. Embattled and embittered, Marland resigned — though he was almost broke and needed the income. Unable to pay the utility bills at their mansion, he and Lydie were forced to move into the lodge/studio on the property.

Marland's story does not end there. He resurfaced as a politician and was elected to the U.S. House of Representatives in 1932. Two years later, he was elected Oklahoma's 10th governor. He and Lydie moved to the capital, Oklahoma City, where

After the merger of Continental Oil Company (Conoco) and Marland Oil in 1929, a new logo topped the Continental Oil headquarters building. While the Denver staff celebrated the transition, opposite, E. W. Marland moved on to politics. As governor, he beautified the state capitol, landscaping the grounds with shrubs transplanted from his estate in Ponca City. In 1936, with two years of his term remaining, Marland made an unsuccessful bid for the U.S. Senate. In 1938, he ran again and was again defeated.

they presided over the rough-and-tumble state like members of royalty, Lydie referred to in one newspaper as a "princess" and Marland photographed often in the company of the famous and wealthy. Occasionally, they would return for a weekend to the Ponca City mansion for fox hunts and polo with such notables as Will Rogers in tow.

As governor, he helped form the Interstate Oil and Compact Commission, a pioneering conservation agency, and he beautified the capitol as he once had Ponca City. But Marland's four years as governor were not without scandal, and he did not win a second term. A shot at a Senate post also failed, as did another oil company he tried to launch. Broke and in ill health in May 1941, he was forced to sell his dream palace to the Discalced Carmelite Fathers for a paltry $66,000.

Over the course of his lifetime, E. W. Marland had won and lost a fortune of $100 million. When he died on October 3, 1941, newspapers called it the end of an era.

In 1929, the Marland Oil Company merged with the Continental Oil Company. The new company adopted the red triangle as its official emblem...with one change. The words "Marland Oils" were painted out on every tank car, filling station, pump station and building and replaced with one word: "Conoco."

Going Global

Pictured is the Ponca City refinery circa 1933. Built in 1918, the refinery was enlarged by E. W. Marland in 1922, increasing its capacity from 8,000 barrels a day to 14,000 barrels. Marland also ordered new 80,000-barrel-capacity storage tanks, increasing overall storage capacity to nearly 2 million barrels. Products were shipped by rail tankcars on tracks that went right through the refinery. By 1928, when Dan Moran took the reins of the company, Ponca City had the largest refinery in the mid-continent region of the United States. The familiar Conoco name — with a new logo incorporating the Marland triangle, opposite — now had the backing of a producing and refining giant.

They say the dust devils were swirling across the Ponca City airport that gray fall day in 1928 when the plane from New York carrying E. W. Marland's successor taxied into the landing area. Hollywood could not have created a more ominous and appropriate backdrop for Dan Moran. Heavy-shouldered and nearly six feet tall, his reputation as a bulldog administrator with an explosive temper had made the rounds of Marland employees. The early reports only scratched the surface.

Legend has it that Moran was born in the shadow of an oil derrick in Cygnet, Ohio. His father and brothers were oil field workers, and he got his first job at age 10 as a messenger for a pipeline company. By the time he arrived in Ponca City, he had served 21 years in the oil industry, "in every branch," he once said, "except legal and the presidency." At age 40, he was eager to tackle the latter.

His first decision was to discharge most of Marland's operating executives, superintendents and managers — people who had grown up with the company alongside E. W. and had gained the respect of fellow employees. "The pink slips just kept coming," recalls Keely Marshall, a Conoco retiree living in Ponca City. "Moran liked to brandish the machete, lopping off people's jobs whenever he felt like it. Back in the 1940s, I was loading scrap metal into a gondola car with another kid. It was August and the temperature was well over 100 degrees [38 degrees centigrade]. My partner told me his safety boots were killing him, and he sat in the shade to pull them off. At precisely that moment, who but Mr. Moran pulls up in his car. 'Nobody sits down on company time,' he says. The kid was fired on the spot."

Such stories are legion. Visit Ponca City, and virtually every old-timer claims to have been a witness to the famous incident in which Moran saw a worker sitting on a stool reading a newspaper and kicked the stool out from under him. "The worker leaped up and punched Moran," says Leland Smith, who concedes he wasn't there. "It turned out he wasn't a Conoco employee at all, but an employee

GOING GLOBAL 71

Ponca City in the 1930s was a fixture on the Oklahoma agricultural circuit, hence the railside mill to the left. The passenger train (a stainless steel "Super Chief" of the Santa Fe Railroad) is being filled with diesel fuel from a Conoco tank truck, circa 1940. Some Conoco tank trucks were unusually large, such as the "Big Berthas" — 6,000-gallon behemoths that caused jaws to drop when they passed. The trucks emptied their precious cargo into underground storage tanks at service stations. Customers' gas tanks were filled by courteous men and women in uniform and hats (bearing the Conoco red triangle, of course) from "modern" gas pumps like the one to the right.

of the phone company installing lines at the refinery!" Moran is said to have been embarrassed not over his mistake, but that the man got in a punch.

These tales today are told with humor. However, another of Moran's capricious acts still rankles retirees. In 1939, Moran had built a recreation building, cafeteria, executive dining room and lounges across the street from the main office building and, to join the two, had built an enclosed walkway over the street. Moran directed that the walkway be partitioned. "He didn't want to walk down the hallway from the offices to the executive dining room and be interfered with by common employees," says James York, a longtime Conoco employee in Ponca City. "Most employees never forgave him for that."

But somewhere beyond the anecdotes is the story of a determined individual who reshaped and guided Conoco through both the Great Depression and the Second World War. When Moran took the wheel in 1928, the company was bleeding red ink, facing debt totaling $6.5 million. The onset of the Great Depression would further exacerbate an already difficult situation. Prospects in the oil industry were poor, given the twin evils of overproduction and soft prices. Moran made it clear that the only way the company could survive was through iron-fisted control of every penny spent.

He took to the task like a military officer of the time, former workers say, gutting their salaries, auctioning surplus equipment, abandoning big exploration plays and selling acres of oil-producing property he felt were marginally profitable. Moran even sounded austere. "I assure you the decisions which have been executed were impelled by stern necessity," he said in 1930. "While necessarily painful to some, [they have] had the most salutary effect upon the organization."

But successful downsizing is not Moran's sole legacy. He set his sights on expanding Marland Oil's modest refinery capacity and marketing depth, an aim supported by his backers on Wall Street. In January 1929, he acquired the Prudential Refining Company, which owned a large refinery in Baltimore but had been steadily losing money. It seemed an unlikely purchase for a company in financial trouble, but Moran was convinced he could turn it around.

Three months later, on April 30, 1929, Marland Oil acquired the Continental Oil Company. Conoco had already drawn the attention of at least two major Wall Street investors,

GOING GLOBAL 73

LINDBERGH FUELS UP
WITH CONOCO

On May 21, 1927, aviator Charles A. Lindbergh completed the first solo flight across the Atlantic, for which he became a well-known hero. What is not well known is whether the famed pilot used Conoco products during his historic flight. There are many references to Lindbergh in Conoco publications from the late 1920s through the early 1960s. Several intimate he may have used Conoco's aviation fuel on his voyage, while others simply note his fondness for the company's wares. Indeed, during his goodwill tour of America following the solo flight, Lindbergh was photographed in Denver and Salt Lake City fueling up with Conoco Bronze. One account even quotes the aviator saying upon landing, "Charles Lindbergh speaking — send the Conoco truck."

Lucky Lindy wasn't the only aviator with a bent for Conoco. An issue of *Conoco* magazine from March 1930 cites another ardent fan — Amelia Earhart.

one of whom was a Morgan partner. The merger was a logical fit and, rumor has it, also satisfied a more personal interest of Mr. Morgan. "J. P. Morgan wanted the merger simply because he was looking for any way to do away with the Marland name," says John Morrow, who retired as Conoco's group senior vice president of finance in the mid-1980s.

Dan Moran liked the match with Continental, figuring the two companies together could solve problems they could not solve separately. Continental needed a steady, inexpensive supply of crude oil, which Marland had aplenty. Meanwhile, Marland needed more marketing outlets for its refined products — Continental's strength. Taking advantage of this consumer visibility, the Marland board authorized the name of the newly formed corporation to be the Continental Oil Company and its headquarters to be in Ponca City.

A worse time could not have been picked for a merger. Both companies (and Prudential) were saddled with debt, a monstrous total of $43 million. Meanwhile, the stock market was gyrating up and (mostly) down. One month after Conoco stock first traded on the New York Stock Exchange on September 15, 1929 (at $34 a share), the stock market crashed, sending shock waves through the capital-starved oil industry.

But Moran was confident that Conoco would outlast the Depression. Despite the company's

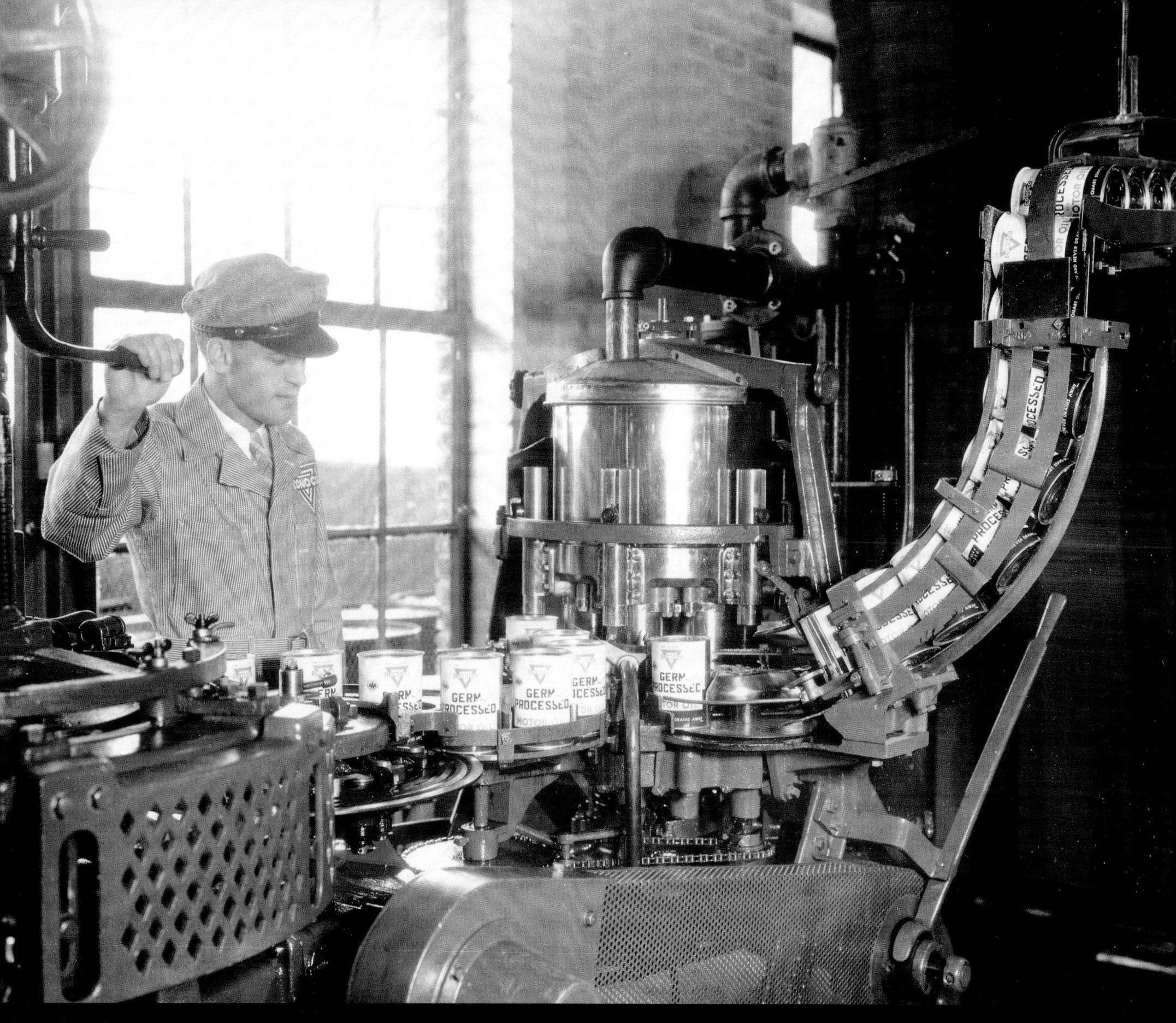
Conoco's Germ-Processed Motor Oil, running through the line at the Ponca City compound and packaging plant in 1934, was considered the company's most valuable asset in the 1930s.

In 1933, Conoco completed the Great Lakes pipeline from Ponca City to Chicago and the Twin Cities — Minneapolis and St. Paul. The 2,100-mile pipeline beat the cost of shipping the products by rail and kept Conoco products competitive in the Midwestern markets of Illinois, Wisconsin and Minnesota. This photo of the pipeline was taken near the crossing of the Mississippi River.

substantial debt, it had much to offer. From Marland Oil had come the Ponca City refinery, the largest in the vicinity of the mid-continent, and the newly acquired Prudential refinery in Baltimore. From Continental had come six smaller refineries in Texas, Oklahoma and the Rocky Mountain states.

Altogether, the companies owned 2,966 wells producing almost 61,000 barrels of crude per day plus a daily refined products output of 46,000 barrels. Its retail outlets dotted 30 states, wholesale markets included buyers in 32 countries, and it held rights to more than two million undeveloped acres. Incredibly, despite the initial thrust of the Depression, the company earned approximately $9 million after taxes in 1929, its first profit in three years.

Moran took aim at strengthening the company's weakest link — product distribution. Conoco had been effectively shut out of the Midwestern market because of the high price of moving its refined products there by railroad. The solution was the Great Lakes pipeline, a 2,100-mile (3,382-kilometer) artery completed in 1933 that connected Ponca City to Chicago and the Twin Cities. From there, Conoco products were trucked to the heavily populated areas of Illinois, Wisconsin and Minnesota in voluminous 6,000-gallon-capacity tank trucks — "Big Berthas," the company called them.

With Conoco's distribution strategy in place, Moran turned his attention to enhancing product quality and advertising. In March 1930, the company introduced the first motor oil in the industry to use an additive that mitigated the effects of engine friction — Conoco's Germ-Processed Motor Oil. "[The lubricant] is the most valuable asset in the possession of this Company," *Conoco* magazine reported at the time.

The germ process was invented by two English scientists who discovered that the reason animal fats are more slippery than petroleum oils is the presence of a free fatty acid. By adding a small quantity of this acid to lubricants, they caused a thin film of oil to build up and remain upon the lubricated surface, thus protecting it. Marland Oil bought the North American rights to the patent in 1923.

After years of top-secret research, Conoco was ready to test its new motor oil against the country's three top-selling brands in contests overseen by the American Automobile Association. Up Pike's Peak, across Death Valley and in the Indianapolis 500 motorcar race, Conoco's Germ-Processed Motor Oil consistently outperformed the competition. "Why use a feebler lubricant, when this test-proven, sturdiest of all motor oils awaits you?" a Conoco advertisement in the *Saturday Evening Post* inquired.

To introduce the product to the public, Moran created the Conoco Germ Process Bus, a "traveling billboard" that toured the country equipped with lantern slides, films and displays touting the new oil. "In smaller towns, we'd park the bus near a Conoco station and play records over the loudspeaker, inviting the public to inspect the bus," recalled Ray Stewart, the bus driver in 1932. "That bright red and green bus was a real attention-getter."

The company also hit the airwaves, sponsoring the *Conoco Show* on the fledgling NBC radio network. The first show was January 16, 1930, from 11:00 to 11:30 p.m. eastern standard time. "We bring to you another in a series of broadcasts by the Conoco Adventurers," announced the (reportedly)

mellow voice of the host, Ralph Wentworth, each night. "The Conoco Singers, Orchestra and Players combine to take you with them on a trip into the romantic and interesting history of Conocoland."

Each broadcast dramatized a historical event from one of Conoco's major market areas, such as the Cherokee Run in Oklahoma or Custer's Last Stand in Montana. The tales were interspersed with music by Conoco's 22-piece orchestra and singing quartet and commercials proclaiming the benefits

Conoco's AAA-rated Germ-Processed Motor Oil, introduced in 1930, successfully tested against the country's three top-selling brands in highly publicized races across Death Valley and in the 1933 Indianapolis 500. Conoco's oil was certified consistently as the best engine-protecting product on the market. It also didn't burn as quickly as other lubricants — hence the extra "Hidden Quart."

GOING GLOBAL 77

THE CONOCO
TRAVEL BUREAU

In 1929, President Dan Moran had an idea to launch a service to help Americans travel the country. In a letter to Conoco executives, Moran outlined his plan for a Conoco Travel Bureau, in which adventurous citizens could "See America First" (some believe Moran may have actually coined the popular phrase in his letter). There were 22 million vehicles in the country at the time, and the travel bureau would provide drivers maps, the schedules of trains and sightseeing buses, and pamphlets about important sites. This trip-planning service would be available at any of the company's 2,000 service stations. All customers had to do was fill out an application providing brief details on their vacation plans. The Conoco Travel Bureau, launched in 1930, would do the rest.

The innovative program was the first of its kind and would prevail for 30 years. During this time, Conoco improved the service, adding marked routes and the Conoco Passport (seen to the right), an official-looking document enclosed in a leatherette case. The passport was the official "introduction to Conocoland."

of Germ-Processed Motor Oil or Conoco's new Ethyl Gasoline, the first anti-knock engine fuel. Broadcast from New York City, the program often featured Broadway stars as guests, including Ginger Rogers, "breaking in with two melody hits," Wentworth chimed.

In 1930, the company's advertising brain trust came up with another clever concept — the Conoco Travel Bureau. The bureau provided free travel information and vacation planning assistance to motorists. "You'd go to your Conoco gas station and tell them you wanted to visit, say, the Grand Canyon," recalls Claudia Mitchell, a former Conoco employee whose father and grandfather also worked for the company. "A few days later, you'd get this packet of information that included a highway map with the best route to take clearly marked, as well as literature on points of interest — and other Conoco stations — along the way."

The material describing the service was contained in a booklet, "Travel Bureau and Passport Service," that was "to be the official introduction and a real

GOING GLOBAL 79

This sticker on your windshield or rear window will identify you as a member of the Conoco Travel Club and assure you of special services at 22,000 Conoco Service Stations and Dealers.

The Conoco Travel Bureau, Denver, Colorado, will gladly furnish you individual trip planning services, including marked road maps, literature, hotel and cottage camp directories, and other helpful information--FREE.

Conoco's Travel Bureau coined the word "Touraide" in 1936 to describe its new map-routing service. Touraide featured spiral-bound booklets with sectionalized maps assembled in the order of a vacationer's trip. Information about scenic attractions along the marked route was provided on the pages opposite the maps. Conoco sent as many as 400,000 Touraides annually to customers in the United States, Canada and Mexico until 1959, when the system of providing marked maps was discontinued.

GOING GLOBAL   81

BUS TRIPS
WITH DAN MORAN

Dan Moran was appointed by J. P. Morgan Jr. to succeed E. W. Marland as head of Marland Oil. He was a colorful character with the stern demeanor and discipline to get the company through the Great Depression. There are endless tales about the famous bus trips in which Moran and troops of executives would board buses and ride hundreds of miles across the country visiting Conoco installations. Though the photo at right shows Moran, camera in hand, during a lighter moment, the on-board employee reviews were notoriously stressful for the people subjected to the "inquisition" tactics of Moran. Moran also was fanatical in his insistence upon neatness. More than one employee was fired on the spot when Moran found tools out of place. He later introduced storage techniques like tool walls with painted outlines indicating where various tools should be hung to avoid accidents or inefficiencies.

passport into Conocoland," early advertising material stated. These "passports" were mailed to service stations from the bureau's headquarters in Denver, site of the company's newest refinery, built in 1931. The travel service continued as a valuable motorist service until the late 1960s.

Conoco employees during the Depression worked the gamut of jobs typical for an integrated oil company. "I was a roustabout for a couple years in the 1930s," says Wayne Glenn, who rose through the ranks to become president of Conoco's Western Hemisphere petroleum division in the late 1970s. "We dug ditches, built pumping units, cut the brush and did all the really grubby work of building an oil field — steam fitting, painting, carpentry, you name it."

Mildred Mock, a file clerk in Conoco's Albuquerque, New Mexico, marketing department during the Depression, later recalled earning less than $10 a week (which included a half-day on Saturday). "Jobs were scarce, and this was a good one," Mock later recalled. Desks during the decade featured "a bowl of water and a sponge for finger wetting, and I remember the company required us always to wear a hat outdoors and turn in old pencil stubs before getting a new pencil." Employees were also issued paperweights — with CLL, the company's ticker symbol, on them — to fend off the wind that arose from open windows in the days before air conditioning.

Safety and loyalty were always rewarded at Conoco. Claudia Mitchell recalls her father's pride as he went to work each day at the Ponca City refinery dressed in his operator's uniform, above. He earned one stripe on his uniform for every five years of service with the company. The typical identification badge above was probably worn by a refinery worker, a practical safety measure at Conoco. The badge at right was awarded to employees to commemorate 40 years of safety.

GOING GLOBAL 83

Bob Zama, a file clerk in Houston in 1935, remembered a six-and-one-half-day workweek. "I finally got smart with a co-worker and we agreed to take off alternate Sundays," he noted years later. Claudia Mitchell recalls her father, an operator in Conoco's light oil division at the Ponca City refinery, wearing a starched khaki uniform to work during the Depression. "Every five years, he'd get a green stripe added to his sleeve showing his years of service," Mitchell says. "And everybody had to wear these incredibly heavy, steel-toed shoes [an early

Conoco Bronze, a higher-octane gasoline, was introduced in 1933 and advertised as "Liquid Lightning." Other ads asserted "Gentlemen prefer Bronze," a send-up of the popular saying, "Gentlemen prefer blondes." The ads were a hit with the driving public, which pumped up record gas sales for Conoco in 1935.

During the Great Depression, President Dan Moran kept the paychecks coming, left. He successfully reduced the company's debt, weeded out what he called "the dead wood" and reintroduced habits of frugality and careful management. To celebrate the company's prosperity during the period, Moran distributed some 5,000 Christmas bonus checks to employees in 1937, below. Worth $770,000 in total, it was one of the largest bonuses bestowed by a U.S. company that year.

indication of Conoco's commitment to safety]."

In their off-hours, Conoco's more athletic employees joined the company's sports teams or went swimming in the Ponca City refinery's pool. Conoco even sponsored a basketball team for female employees — the "Ethyls."

Meanwhile, the company continually tested new products, such as its higher-octane Bronze gasoline, unveiled in 1933 and advertised as "Liquid Lightning." More money was poured into advertising. In June 1935, Conoco released a series of award-winning print advertisements featuring an illustration of an elegant woman driving a convertible, her hair streaming behind her. Above were three words: "Gentlemen prefer Bronze." Conoco's gasoline sales escalated, up 17.7 percent in 1935 alone, a year in which it reported the greatest volume of business in its 60-year history.

By 1937, Moran had eliminated the company's $43 million debt. The Depression was effectively over for Conoco's employees. On the steps of the Ponca City headquarters that Christmas, he distributed to employees 5,000 bonus checks worth $770,000 in all — one of the biggest bonuses bestowed by any company in the United States that year. "I remember we were all gathered out in front

GOING GLOBAL

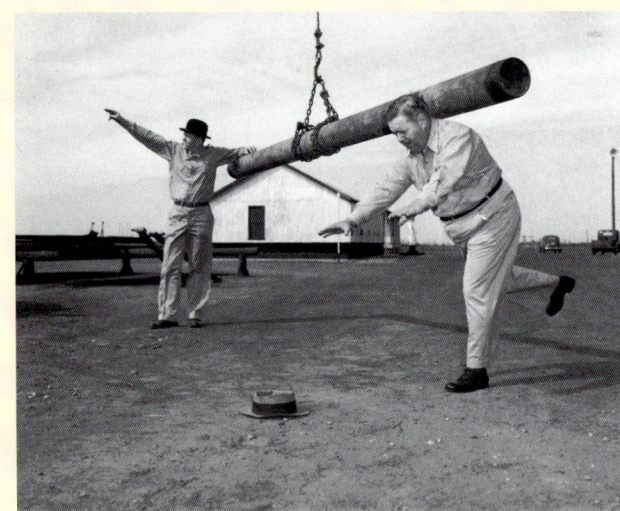

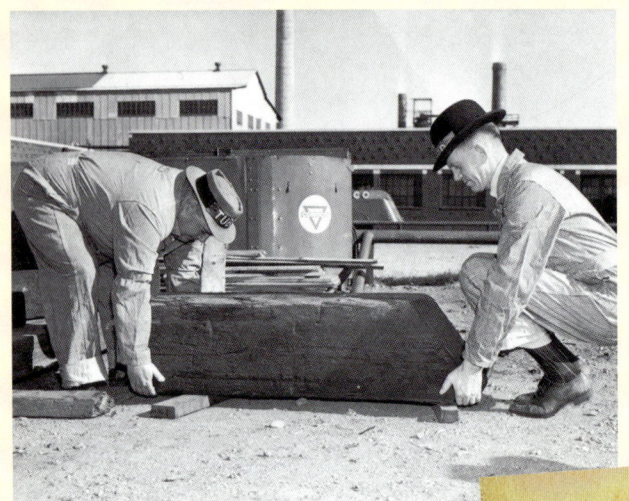

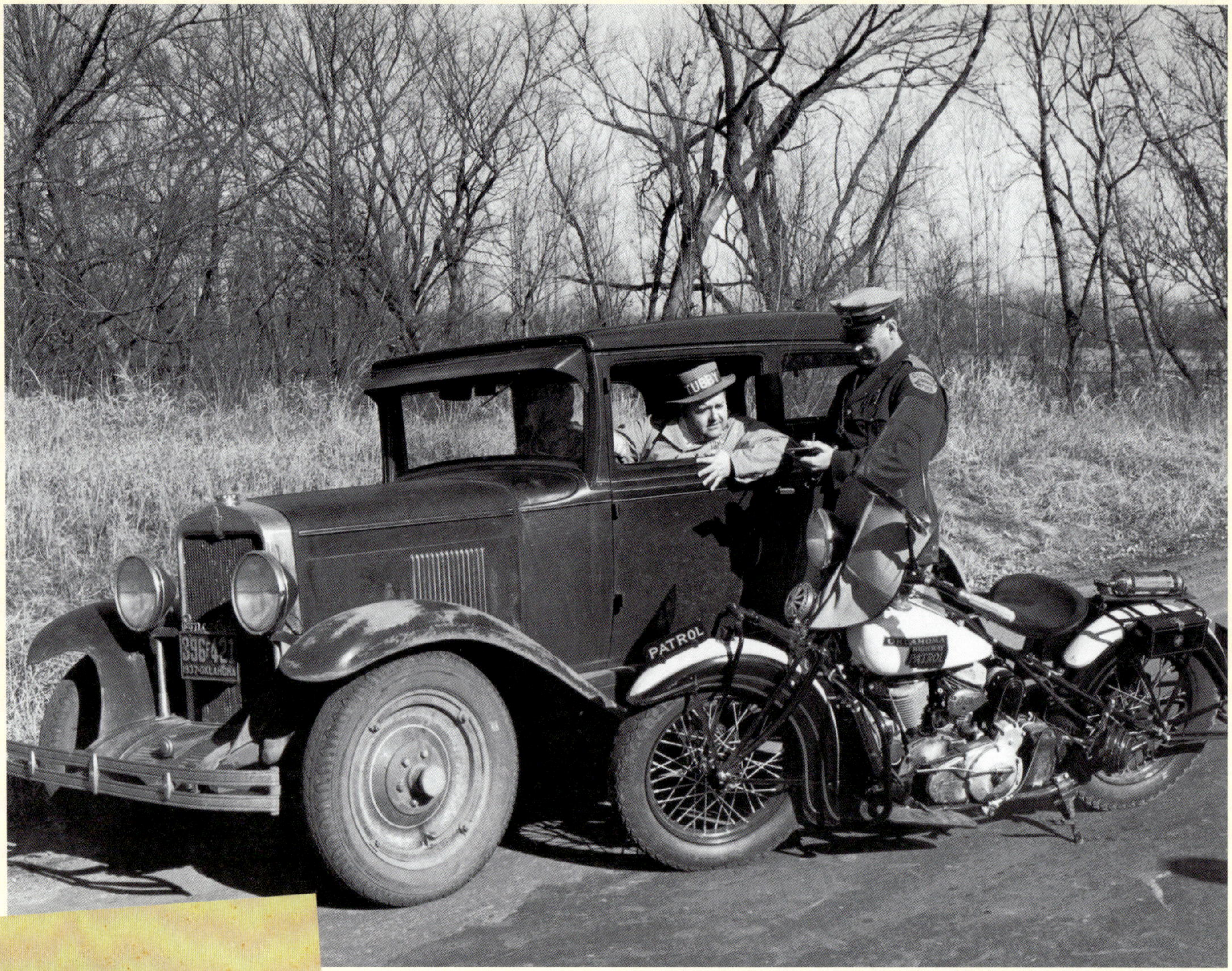

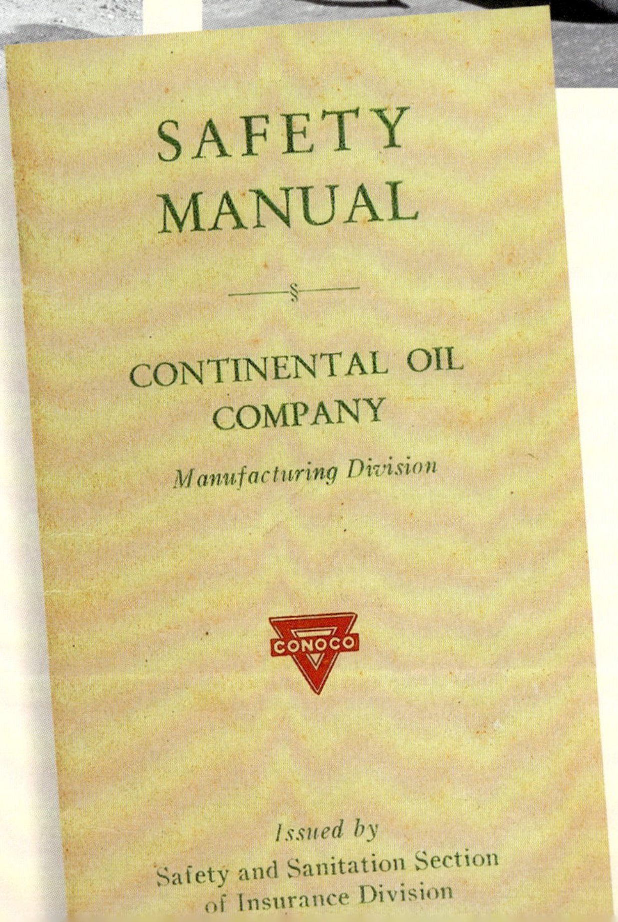
SAFETY
MANUAL

CONTINENTAL OIL
COMPANY

Manufacturing Division

CONOCO

Issued by
Safety and Sanitation Section
of Insurance Division

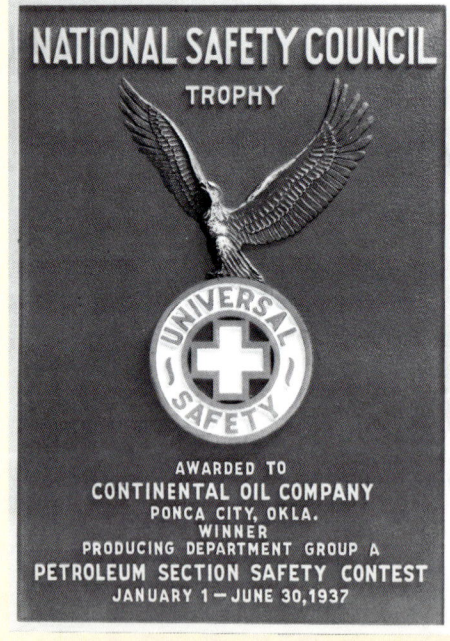

THE CONOCO SAFETY RECORD

Pictured opposite are the infamous Tubby and Red, a.k.a. Herbert "Tubby" Wilbur and "Red" Steinfort. They were recruited from Conoco's production accounting department in 1935 to be photographed for the *Profit Taker*, a Conoco publication focused on safety ("banishing the profit takers — fire and accidents").

The Tubby and Red photos presented a humorous take on a very serious issue — safety. The idea was reportedly hatched by Dr. Walter Miller, vice president for manufacturing in the 1930s. Miller advocated such safety measures as proper lifting techniques, eye safety and a strict "no horseplay" rule, topics that Tubby and Red illustrated in their comedic skits.

The focus on safety had the desired effect. Conoco's excellent safety record was recognized by the National Safety Council in 1937. Conocoans were among the NSC's Trophy Award winners. Conoco ran a publicity campaign congratulating the winners with a "Hats Off" salute.

GOING GLOBAL 87

of the main office building and, one by one, were handed our bonuses," recalls Wanda Lee Fisher Jones, a third-generation Conoco employee who retired in 1972. "The newspapers were there, flashbulbs were popping, and everybody was extremely excited." Moran knew a good photo opportunity when he saw one.

The day was to be Dan Moran's finest. His Spartan and sometimes severe methods forged a scrappy Oklahoma oil company. "What Moran sought to build was not a great national company, but a tightly knit regional company, on the fringe of greatness, playing its own special game," *Fortune* magazine commented in April 1961. "And in this he succeeded admirably, [despite] his tyrannical one man show."

Moran would make several pivotal decisions before ill health forced his resignation in 1947, among them Conoco's "Louisiana Purchase" — increasing Conoco's holdings of the oil-rich land in Louisiana to 88,000 acres by the end of the 1930s. A pipeline connecting the southern Louisiana fields to the Gulf port city of Lake Charles, Louisiana, transported oil to a $4.5 million modern refinery Conoco completed in July 1941.

A new lubricant, Conoco Nth Motor Oil, was placed on the market that year, tested by the same rigorous standards as the company's successful germ-processed lubricant. At the same time, scientific advances in the development of synthetic

Although he is remembered more for his explosive temper and gruff demeanor, Dan Moran had his lighter side. He once invited company executives to a hunting party at his beloved Mo-Ranch, near Kerrville, Texas. Included with the invitation was a paper bag with instructions to remember to take scraps home to the family dog. Another invitation contained shotgun shells! Although Moran's legacy is forever wedded to his image as a cigar-chomping authoritarian, his extreme ways did navigate Conoco over the shoals of the Great Depression. As one Conocoan recalled, "You need a tough man to survive hard times."

The safety programs initiated in the early years of the company's history became a critical aspect of the Conoco culture. The company's fire safety programs of today, for instance, have their origin in Conoco's fire patrol of the 1930s. The fire patrol was essentially an on-site fire department with its own trucks, hoses and fire chief.

Conoco's Houston offices until the early 1940s. In 1949, Conoco's new president, Leonard F. McCollum, moved the company's headquarters from Ponca City to Houston — the center of the U.S. petroleum industry. New offices were obtained at the Sterling Building on Texas and Fannin Streets.

products from petroleum were captivating the oil industry and Conoco especially. Most early efforts in this area were failures, but when the United States entered the war following the bombing of Pearl Harbor in 1941, the need for wartime supplies focused new attention on petrochemicals — launching, in effect, a new industry.

High-octane aviation fuel was a wartime necessity and an area of the war effort in which Conoco's capacity for innovation sped Allied victory. Conoco, along with others, pioneered the alkylation process, which produced the key blend stock of the fuel needed for high-altitude and high-performance war planes. Both the Lake Charles and the Ponca City refineries produced the fuel, and at Ponca City, Conoco also operated a separate refinery on behalf of the U.S. government to manufacture 100-octane

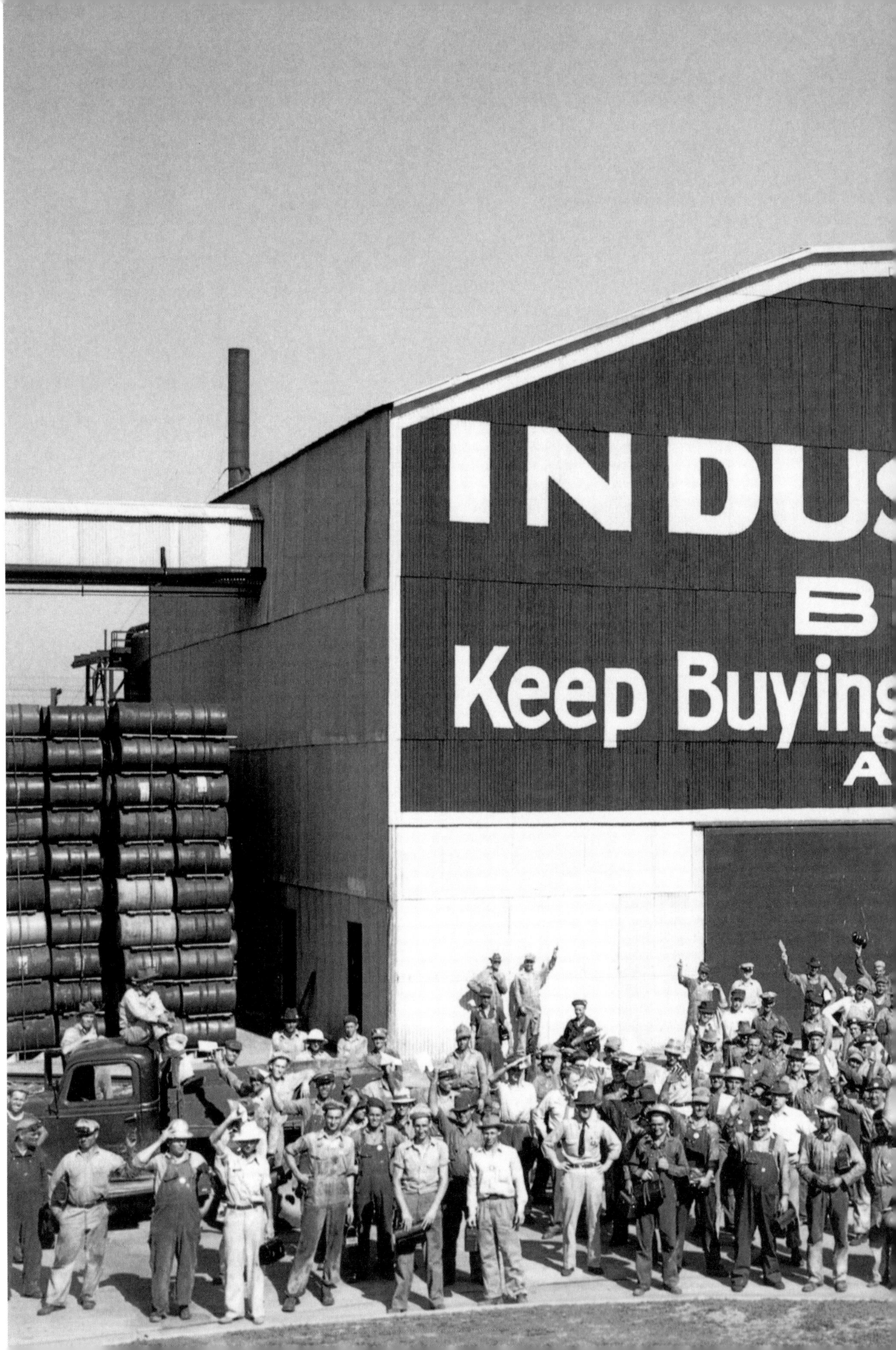

Thanks to its long aviation history, Ponca City was chosen as one of the six sites at which British Royal Air Force pilots trained during World War II. The Darr Pilot Training Center trained more than 1,200 RAF pilots and 250 American pilots from August 1941 through April 1944. Researchers from Conoco and other oil companies developed a high-performance, high-altitude fuel for Allied planes, and Conoco operated a separate refinery in Ponca City to produce government orders of 100-octane aviation fuel, which boosted aircraft performance. Conoco had a close connection to the Darr school, providing aviation fuel for the planes, as seen on pages 94 and 95 following. It wasn't all work, however. Opposite, Conoco "junior associates," (children of employees) join the cadets at an annual celebration for the pilots, the Wings Dinner and Dance, hosted by Conoco.

GOING GLOBAL 93

During World War II, denim-clad women filled the jobs of men fighting overseas. In 1943, 15 women joined the company as the first female refinery workers in its history. Their ranks exploded the following year. Some women even received formal "Discharge Papers" from Conoco in recognition for their wartime service.

aviation gasoline. The new fuel provided Allied warplanes with a 20 percent boost in power over previous fuels, giving the faster, higher-flying planes a decided edge over enemy aircraft. Demand also grew for nylon and other materials that could be produced from petrochemicals to manufacture parachutes, helmets and parts for weapons.

Many Conoco employees shipped overseas to give their all for the war effort. As men left the refinery to do their duty, women replaced them — more than 1,000 in all. "The work we were doing, particularly for Allied airplanes, was very secretive, and there were watchmen patrolling outside with dogs," says Irene B. Senseman, who worked in the company's laboratory as a motor oil tester during the war. "I remember a lot of the girls lived together but rarely saw each other because of the different shifts. And we all wore blue denim overalls emblazoned with the red Conoco triangle and kerchiefs to keep our hair in place."

The manpower shortage also required ingenuity of a more prosaic sort. To mow the grass growing around the company's 786-acre oil storage tank farm, Conoco spent $7,500 in 1942 on an outside firm. When the price jumped to a prohibitive $10,000 the following year, the ever-frugal Moran hit on a novel money-saving idea: He bought 2,000 sheep and let them do the job! After the war, the sheep were shipped off to market.

Conoco grew very little during the war years —

Several products invented by Conoco scientists and engineers in the late 1930s entered the marketplace, including the first anti-squeak crayon. The company built on its reputation and retail network to successfully market new products geared to household and gardening needs. Conoco's Germ-Processed Motor Oil came in a lighter version for lawnmowers and other small machines. The company's household products ranged from spot removers to floor waxes and were sold at Conoco service stations.

barely 10 percent in total assets from 1941 until V-J Day. The first full peacetime year, 1946, however, was a point of takeoff. "The industry, suddenly unchained from government restrictions and wartime limitations, rebounded with competitive fervor," company historians later commented.

Over the course of the next 25 years, a new leader would guide Conoco. As the company's president, chief executive officer and eventual chairman, he would expand Conoco domestically and, brick by brick, build it into an integrated, international oil company. His name was Leonard F. McCollum, but everyone knew him simply as Mr. Mac.

L F. McCollum's first executive order as Conoco's new president in 1947 would forever endear him to employees. He literally and figuratively tore down the vilified partition that Dan Moran had erected years before. The order sent a clear message — this president would not be a remote figure cloistered in an ivory tower. Where Moran was withdrawn, autocratic and parsimonious, McCollum was open, flexible and quite eager to spend money — especially on exploration.

Tireless and trim with crisp, good-humored eyes and an almost courtly manner, the enthusiastic McCollum brought a new optimism to the company. "He was the absolute antithesis of Moran," says Bill Thomas, a graphic artist at Conoco at the time of

Land thickly dotted with oil storage tanks is rarely thought of as a habitat for sheep. But on Conoco's tank farm at Ponca City during the war, more than 2,000 sheep spent their pre-market days nibbling the grass growing around the tanks. This woolly idea had a purpose: Due to the manpower shortages caused by the war, groundskeeping costs at the 786-acre tank farm had skyrocketed to $10,000 a year. So Conoco turned instead to this unique laborsaving strategy. After the war, the flock was sent to market, and the company resumed its customary maintenance procedures.

Leonard F. McCollum, a Tennessee farm boy who became Conoco president upon the resignation of Dan Moran, was known throughout the oil industry as Mr. Mac. Where Moran was autocratic and tight-fisted, McCollum was flexible and eager to invest in the company's growth, particularly in international exploration. His first corporate act was to remove the wall Moran erected to separate Conoco employees from executives in the walkway leading to the company's cafeteria in Ponca City. The message was clear: Mr. Mac believed in the virtues of teamwork, frank communication and collaboration.

McCollum's appointment to the top job. "He was friendly, good-natured and engaging. Everyone here called him Mr. Mac, a nickname he encouraged because of its informality. We said it, though, with respect."

In his first year on the job, McCollum made it a point to visit every Conoco location and talk to employees, and he later remembered them by name. "He was gracious," says Keely Marshall. "He established a quick and easy rapport with you, no matter where you worked in the company. And he never had a harsh word for anyone. As a result, people were eager to please him."

Born on a Tennessee farm, McCollum liked to say he stumbled into the oil business while trying to meet the science requirements at the University of Texas, where he had enrolled to study journalism. Looking for the easiest course he could find, he picked geology, fell in love with it and eventually earned a B.A. in the field in 1925. He went to work as an oil scout for the Humble Oil Company in Texas and then was transferred to another Standard Oil affiliate, Carter Oil, in Tulsa, Oklahoma, as its exploration manager. "Those were wild and woolly days," he later recalled. "Oil was flowing, speculators were making fortunes, and oilmen were going from poverty to wealth overnight."

McCollum's rise was swift. He was named president of Carter Oil in 1941 at age 39, becoming one of the youngest company heads in the oil industry, and there seemed a strong likelihood he might one day go to the very top of Standard Oil. But when Dan Moran took ill in 1947, Conoco made McCollum a "very compelling" offer, he later said. He signed on as president in August of that year.

Moran had left Conoco in sound financial shape, with total assets of just over $209 million and no debt at the end of 1947. "Of course, with the expansionary plans I had, that debt-free situation didn't last long," McCollum noted with a laugh in a 1986 interview. "Moran's methods served the company well during the Depression, but it was time for building — not caretaking."

Old-timers recall that on his first day in office, McCollum inquired about the location of Conoco's research department. Informed that Moran had dismantled it, McCollum insisted another be built immediately. Four researchers — all Ph.D.s — were hired to oversee the development of a new R&D department. By the end of 1949, they had assembled a staff of 128 scientists and technicians. Research was conducted on every facet related to the oil business, much of it in the production area. One year later, to mark the 75th anniversary of the company, construction began on a new $2.25 million research laboratory in Ponca City.

"McCollum was always tremendously interested in research, and he created an atmosphere in which new things could be tried," said Dr. Marion "Slim" Sharrah, one of those first four researchers. "A lot of our major successes — the Vibroseis seismic system, our new Super Motor Oil [introduced in 1950] and our synthetic detergent work — can be traced back to his encouragement."

The young chief executive was bursting with ideas. He created a coordinating and planning department headed by Serge Jurenev, a respected

Conoco's 75th anniversary celebration in 1950 carried a groundbreaking theme that was more than metaphorical: Conoco had just begun digging the foundation for a $2.25 million research laboratory in Ponca City.

GOING GLOBAL 101

E. W. Marland established the oil industry's first research division in the late 1910s, staffing it with more than a dozen geologists and physicists. The division paved the way for a series of technological advancements, including the use of the core drill to search for oil deposits. As the years progressed, Conoco continued to invest in more modern research facilities. The $2.25 million research lab built by L. F. McCollum in 1950, below, was the site of the development of the Vibroseis seismic system of exploration, the company's new Super Motor Oil and its synthetic detergent products of the 1950s and 1960s. At right is a laboratory technician from the 1960s conducting research in the Ponca City facility.

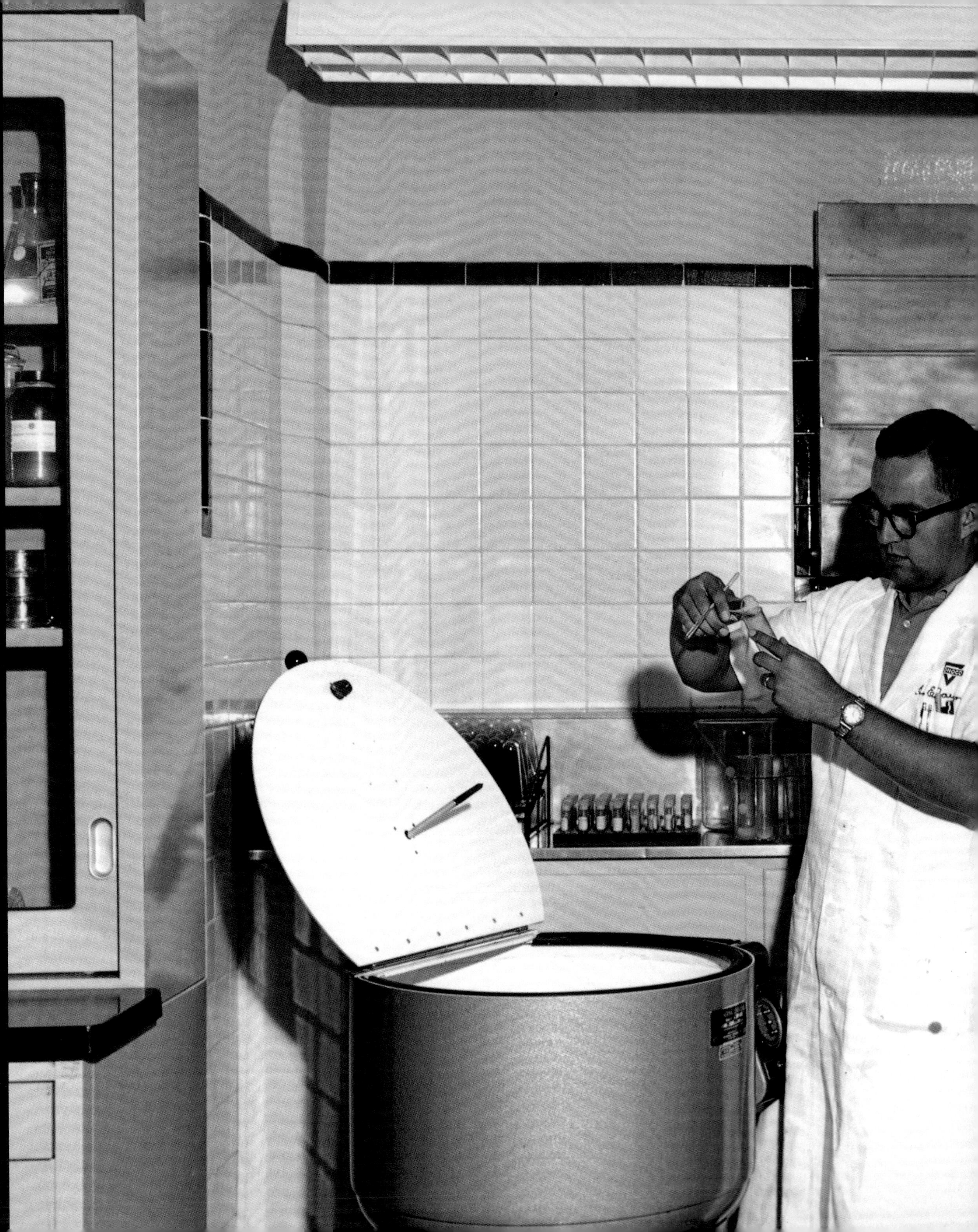

economist. The department analyzed potential acquisitions and the company's asset base. To improve Conoco's refined products capacity, McCollum launched a refinery construction and modernization program in the early 1950s. The Ponca City, Lake Charles and Denver refineries were enlarged, the Baltimore refinery was converted to produce synthetic detergent intermediates, and a new refinery was built at Billings, Montana. The Lake Charles refinery's capacity alone was tripled to 40,000 barrels a day, and a plant to manufacture carbon black, another petrochemical synthetic, was constructed adjacent to it.

McCollum also reorganized the company's operating strategy. "Under Moran, word came down stenciled in concrete what you had to do and when it had to be done," says John Morrow. "Mr. Mac recognized that never before had managers been asked to do their own thinking." In 1948, McCollum organized the company's first management meeting in Colorado Springs. He moved Conoco's corporate headquarters from Ponca City to Houston in 1949 to be closer, he said, to a financial and petroleum center. The company set up offices at the Sterling Building on Texas and Fannin Streets. A year later, McCollum restructured the company along decentralized lines, creating five regional headquarters — in Denver, Houston, Oklahoma City, Los Angeles and Fort Worth. "He had complete faith in his regional superintendents to

GOING GLOBAL

Conoco was the first American oil company to employ seismography as an exploration tool. In 1921, E. W. Marland assembled the first refraction seismography crew in the U.S. Marland hired the German scientists who had developed this technology to work alongside the company's geological staff. In the 1940s and 1950s, Conoco deployed several seismic crews to map potential structures. Dynamite provided the source of energy directed into the earth, and the returning waves were recorded to produce the basis of geological maps.

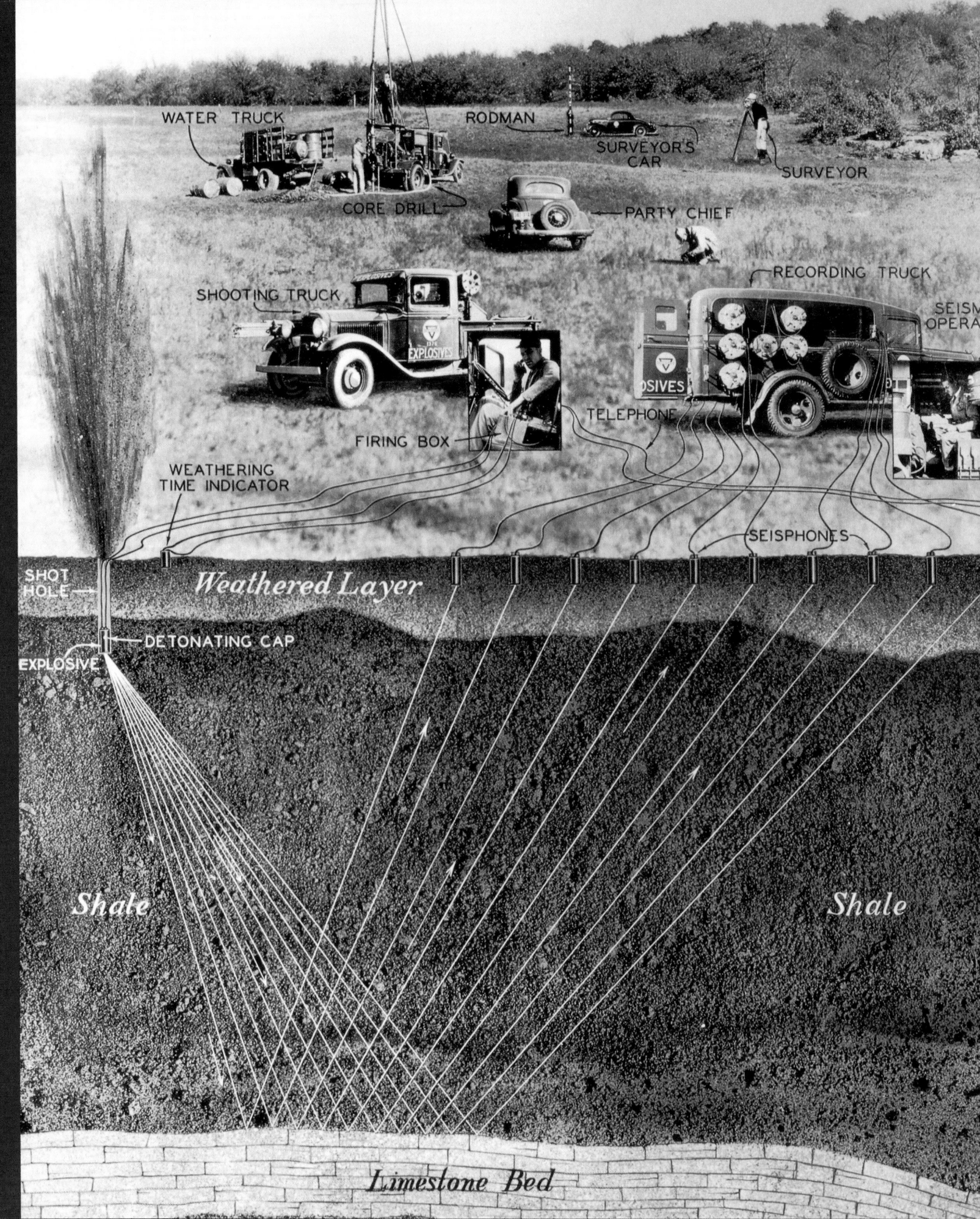

effectively oversee their areas of responsibility," recalls Constantine "Dino" Nicandros, a McCollum protégé who later became president and chief executive officer of Conoco.

McCollum made it clear that he expected his managers to develop new projects and that he was willing to spend money on good ideas. No subject captivated him more than the hunt for oil. "He had a firm belief in exploration and the creation of wealth from discovery," Nicandros says. "Such a strategy, and it was the right one at the time, requires a willingness to spend quite a bit of money." To finance Conoco's exploration-related debts and future expansion plans, the company issued $100 million in bonds in 1954.

McCollum brought in a visionary geologist, Dr. Ira Cram, and named him vice president of exploration, charged with expanding the company's horizons. Cram and McCollum collaborated on new directions for Conoco's exploration effort, with emphasis on significantly expanded programs in the Gulf of Mexico and pioneering efforts internationally.

Conoco had begun to explore offshore for oil in the Gulf of Mexico under Moran in the mid-1940s. Ingeniously, the company retrofitted flat-bottomed boats used in World War II beach invasions with seismic equipment to locate promising geologic structures. The early results were promising, and in 1946, Conoco organized and managed a syndicate

for offshore exploration in the Gulf called CATC — the members being Conoco, Atlantic Refining, Tidewater Oil and Cities Service. CATC drilled its first producing well in the Gulf in 1953.

Off the coast of California, Conoco was the operator for another partnership formed in 1953 to explore for oil. The CUSS Group, which included Conoco, Union, Shell and Superior, drilled not from a rig planted solidly on the seafloor — the customary method — but from a floating vessel at anchor. "We had bought some leases in 190 feet [58 meters] of water, but there were no rigs capable of drilling out there," recalled Vic Eissler, a Conoco engineer working the Gulf at the time.

GOING GLOBAL 105

This photo, dating from 1960, depicts the digging of a labyrinthine cavern below the Ponca City refinery to store propane, butane and other liquefied petroleum gases. The cavern is still in use today, storing about 275,000 barrels of LPGs, which are processed by the refinery above and then sold commercially. In the "propane cave," as Conocoans call it, lives an old Caterpillar bulldozer dating from the year of its building. The earth-moving machine had to be disassembled to fit into the cave, where it was then reassembled. Unfortunately, it was too costly to repeat the process once the cave was completed, so the dozer remains in the cave, a ghostly sentinel amid the gases.

"So we integrated a drilling rig onto a deck section onshore and towed it out." The drilling platform was constructed amidships on a surplus U.S. Navy patrol boat that CUSS acquired and christened *Submarex*. A second ship, *CUSS I*, was similarly outfitted. These were the first ships in history to make full-scale drilling from a vessel at sea practical.

When oil began pouring out of western Canada's Leduc field in Alberta in 1947, McCollum reactivated Conoco's dormant Hudson's Bay Oil and Gas Company, familiarly called H-BOG, the operating company E. W. Marland had created in 1926 over tea and crumpets. H-BOG had been inactive since 1934, due to the effects of the Depression, but Conoco still had four years left on its leases. McCollum pumped capital into H-BOG's exploration activities on some 6 million acres under option in Alberta, Manitoba and Saskatchewan. As a result, production shot up at an annual rate of about 35 percent.

The company also began serious development of its natural gas and liquefied petroleum businesses. Prior to 1950, gas was not all that important to sales. The 1930 annual report noted natural gas and gas facilities income of $5.7 million, or about 2.5 percent of total revenues. The 1940 annual report doesn't even report natural gas production. But by 1950, the company's 75th anniversary, things had begun to change — natural gas and liquefied petroleum gases (LPG) merited an entire page. Deliveries of gas that year had grown tenfold from the 1930 level to 107 billion cubic feet, and gas liquids measured 3.3 million barrels.

In 1955, Conoco and the Union Stock Yard Company formed the Constock Liquid Methane Corporation. Constock was to develop the technology to liquefy natural gas (LNG) at minus 258 degrees Fahrenheit (minus 161 degrees centigrade) and to ship, store and re-gasify the liquid methane for sales into gas distribution systems. Much of this development was overseen by Ed Batson and John Murphy. "The Union Stock Yard group was interested in this process in part because it could gain free refrigeration for its meat operations as the gas cooled during the re-gasification process," says Mike Stinson, senior vice president of government affairs, corporate strategy and communications. Shortly thereafter, Constock developed a test ship, the *Methane Pioneer*, which shipped liquid methane from Conoco's Lake Charles liquefaction plant to a re-gasification facility at Canvey Island, England, for markets in that country. A new international trade in liquefied natural gas was born.

Having stepped up Conoco's exploration activities in North America, both onshore and offshore, McCollum cast his eyes farther abroad. At the time, Conoco and other U.S. oil companies were threatened by the increasing importation of less-expensive foreign oil, which was then unrestricted. To protect

GOING GLOBAL

Conoco's position at home and hedge against the importers, McCollum went overseas. "I used outside consultants to tell us what were the most likely places in the world to find oil and hired the best international people I could," he later explained. Howard Hinson, vice president of exploration worldwide under McCollum, recalls, "He gave me only one mandate: 'Make Conoco a sizable foreign producer.' He never second-guessed me."

Conoco organized a consortium in 1948 with Ohio Oil and Amerada Petroleum Corp. — called Conorada — to explore for oil outside North America. "We had a blank check as far as international exploration was concerned," says George McAteer, a Conoco retiree who early in his career handled the administrative duties for the company's exploration teams. "Mr. Mac was an explorer himself, you see, and he couldn't encourage us enough."

Conorada's first exploration activities were a series of surface studies in South America and Africa. (Ironically, one of Conorada's first target areas, Venezuela, would become the site of a major exploratory effort more than 30 years later.) In late 1953, the group recommended large tracts in Egypt's great western desert. Concessions were granted to Conoco, along with Marathon, Cities Service and Richfield. But after four years, nine dry holes and $28 million in expenses, Conoco pulled out. McCollum was unperturbed. "You must have the audacity to...take a big bite," he told *Fortune* magazine in 1964.

Conoco had not completely given up on Egypt. In 1954, it formed a subsidiary, the Egyptian-American Oil Company, which later was granted two concessions by the Egyptian government. The concessions covered 88,000 square miles (227,920 square kilometers) along the Mediterranean Sea, from the Nile to the Libyan border. Plans to drill a test well were canceled in 1956, however, when hostilities in the Middle East erupted and Conoco's employees were forced to evacuate.

Still, the company persisted. In 1955, Conoco subsidiaries were among the oil companies granted concessions across the northern part of the kingdom of Libya. "The country had no petroleum laws at the time," recalls Brooks Buxton, president of Conoco Arabia Inc., who was stationed in Libya from 1976 to 1983. At the invitation of the country's prime minister, Conorada and others were asked to suggest draft regulations. Conocoans Bernie Braly and Charles Bell were instrumental in drafting the law that was passed in 1955 granting the concessions.

"Our partners did not want too much acreage on the first round," Hinson recalls. "McCollum eventually succeeded in persuading them to go along with us and take additional acreage. In fact, much of the big production Conoco eventually saw came out of that second round of acreage." Conoco's leases included a large section of land near the Mediterranean Coast that other oil companies had eschewed. Drilling began in earnest in 1958. Although the first well drilled was productive, it was noncommercial. Finally, in 1959, a major oil deposit was discovered at Dahra field. It eventually yielded 100 producing wells with an output of

Libya was the site of Conoco's first overseas oil development. Conoco's unsuccessful exploration program in the western desert of Egypt pointed to Libya as a potentially rich petroleum province. Conorada, a partnership of Conoco, Ohio Oil Co. and Amerada, discovered the giant Dhara field in 1959. They formed Oasis Oil Co. to operate the new field. Oasis constructed a pipeline from the Saharan desert to the Mediterranean Coast to export the oil. The Es Sider pipeline was completed in 1962, and the event was marked by a celebration attended by Libyan dignitaries and representatives from the oil companies, opposite. Dhara, pictured at left with a lone drilling rig, was joined by other fields, eventually producing more than a million barrels of oil per day.

GOING GLOBAL 109

120,000 barrels daily. Conoco's great overseas gamble had paid off at last.

"Had we not struck in Libya, Conoco may have cut down on foreign exploration altogether," Hinson asserts. "Libya was the largest single jump in production that Conoco has ever had. It provided us with both the cash and experience to become an international player."

To operate these oil properties, the partners formed the Oasis Oil Company of Libya. Over the course of the next year, Oasis would build a port and terminal facilities on the Mediterranean Sea at Es Sider, about 87 miles (140 kilometers) from the Dahra field. A pipeline connecting the field to the port was completed in early 1962, and oil began to flow for shipment that May. Total production at Oasis would eventually exceed one million barrels per day.

Conoco's growing Middle East exploration and production activities during the 1950s and 1960s led to an expansion of the company's tanker fleet delivering oil from that region to Europe, the United States and Asia. Conoco's crude oil tanker history dates back to at least 1929, when the company acquired a fleet of six tankers and a seagoing barge, the *Conoco No. 1*, primarily for deliveries at ports on the East and Gulf Coasts of the United States, as Conoco was then primarily a U.S.-based company.

By 1962, Conoco had a foundation for decades of international exploration. McCollum's midsize regional oil company had ventured into the global market, where only the international petroleum giants had dared to tread. And it had struck it big.

Bolstered by its success in Libya and prodded by McCollum's pledge to make Conoco the "best damn oil finder in the business," the company sent seismographic crews roaming distant parts of the globe in search of oil.

An international exploration and production department in New York City was established in 1958 and enlarged in 1960 to oversee Conoco's foreign exploration efforts. "Our responsibilities were to acquire overseas interests external to our Conorada activity," recalls Dick Hittle, one of three Conoco executives who manned the original office.

DOODLEBUGGERS

More than 150 expert seismen — or "doodlebuggers" as they called themselves — worked year-round for Conoco on the company's seismographic crews in the 1940s and 1950s, scouring a million acres annually looking for oil. When out in the field, these "members of the underground" — another identifying nickname — would set up offices in the nearest town to send the seis data directly to the Ponca City lab.

A typical "seis" crew included 15 doodlebuggers; a recording truck; two "shooting" trucks; a surveyor's truck; a drill truck with a small rotary drill and several hundred feet of casing, drill pipe and tamping poles; and one or two water trucks.

The nomadic work was tedious, exhausting and — as any seisman will tell you — enormous fun. "Every day was like the Fourth of July!" one doodlebugger recalled in the *Red Triangle*, a company publication. Evidently, he especially liked the dynamite part of the job. Boom!

VIBROSEIS

Why detonate dynamite when hydraulically produced vibrations can yield the same result? Back in 1953, this idea intrigued Bill Doty and John Crawford, Conoco geophysical researchers in Ponca City. An article in a scientific journal described the process of "correlation," in which coded electrical signals could be separated from high-level noise signals. Doty tossed the article on Crawford's desk. "Here's something we can adapt to seismic oil exploration," he said. They spent the next several months developing and building ground vibrators and measuring the data points produced. After many initial setbacks, on July 3, 1953, a pilot test they conducted in Orlando, Oklahoma, performed in accordance with customary seismic readings of the same substructure.

The geophysicists patented their embryonic technology as "Vibroseis." Today, more than half of the seismic exploration crews in the world still use the system. In 1991, Doty and Crawford were awarded DuPont's prestigious Lavoisier Medal for Technical Achievement.

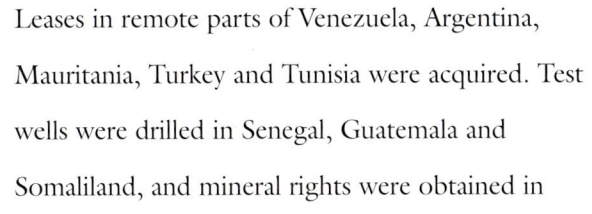

Leases in remote parts of Venezuela, Argentina, Mauritania, Turkey and Tunisia were acquired. Test wells were drilled in Senegal, Guatemala and Somaliland, and mineral rights were obtained in Queensland, Australia, and the Republic of Ireland. The international department moved quickly. By 1961, Conoco was prowling for oil in 21 countries.

Members of the "seis" crews in the U.S. and abroad called themselves "doodlebuggers." Back home, the crews ranged the central, southern and western regions of the U.S. looking for good drilling prospects. "We traversed the country looking for oil, moving our families every four months or so," recalls George McAteer, a doodlebugger in the early 1960s. "We'd live in furnished apartments and be able to move out in two hours. It was a great way to see the country."

The crews used a new geophysical prospecting tool invented in 1956 — Vibroseis. Conoco retiree John Crawford, one of the tool's inventors and later a recipient of DuPont's prestigious Lavoisier Medal for his efforts, explained how it works. "Vibroseis basically involves a mechanical means of generating low-frequency sound waves in place of the explosive detonations that were typically used to generate

ROBERT STOLT

You know the type — the kid who just can't control his curiosity and has to take things apart. "That was me," says Bob Stolt, Conoco senior research fellow in exploration production technology. His breakthrough imaging technique yielded a more accurate definition of physical properties of rock layers, changing the course of oil exploration. The Stolt Migration Process earned the humble scientist the prized DuPont Lavoisier Award in 1998. Stolt modestly claims he was just in the right place at the right time. "Conoco creates an environment that nurtures creativity and allows you to express it," he says. Stolt is still at work on imaging problems. "We're developing ways to fill in the gaps in the seismic data. By using models, we can extrapolate what's not revealed through seismic tests." It sounds quite complex, but Stolt admits to being a mere human being. "I bet if I took a watch apart today, I still couldn't put it back together."

To gain downstream presence in Europe and provide outlets for its Libyan oil, Conoco purchased Jet Petroleum Ltd. in the United Kingdom in 1961. With more than 400 retail outlets at the time, Jet had made a name for itself as a seller of low-price gasoline. Conoco had also acquired about 100 Seca stations in Belgium and Luxembourg and more than 400 Sopi stations in West Germany and Austria.

waves in seismic work at the time," says Crawford. "The tool eventually became standard fare in the oil industry, and we generated significant revenue by licensing it."

The funding of Conoco's heightened North American exploration efforts through the 1950s paid off in 1960 — crude oil production surged to 156,000 barrels a day, a company record at the time. New refineries in Artesia, New Mexico, and Wrenshall, Minnesota, were acquired to help process the rivers of crude that flowed through the company's expanding arterial system, its pipelines. In the mid-1950s, Conoco built the Yellowstone pipeline (the first pipeline to cross the Continental Divide) and was a member of the Platte Pipe Line Company, which built a 1,080-mile (1,738-kilometer) pipeline from Wyoming to Illinois. In 1960, the company began building the Glacier pipeline, connecting oil fields in Canada and Montana to its Billings refinery. Conoco pushed to double the capacity of the Great Lakes pipeline from 1950 to 1960. Other pipelines Conoco helped build or expand during the period included the Pioneer, Cherokee and Colonial. Much of the financing for these pipelines was supported by throughput agreements and innovative, off-balance-sheet debt instruments.

These improvements in Conoco's domestic production, refinery capacity and distribution channels were complemented by an emboldened marketing presence. McCollum acquired three regional oil companies in 1958 — Kendall Oil, Coastal Oil and Western Oil and Fuel. Kendall marketed gasoline in the South under the Kayo name through some 200 retail outlets. Western operated in the Midwest through some 350 retail stations, while Coastal distributed heating oils in the New York-New Jersey area and owned or controlled terminals with more than one million barrels of storage capacity.

Kendall and Western would play important roles in Conoco's evolution. Kendall's Kayo outlets, for example, introduced Conoco to the low-price, lower-service retail strategy, a precursor of today's

self-service stations. "Kayo sells gas at the lowest possible price, with hardly a dab at the windshield," *Fortune* reported in 1964. Western's retail stations, on the other hand, included a small convenience store, giving Conoco its initial experience selling products other than gasoline, motor oil and so-called TBA (tires, batteries and accessories, a line it entered in 1950). "We used gasoline as a loss leader to bring people into our stores," says Maria Joy, a Conoco retiree who started her career in 1962 with Western in Minneapolis. "Eventually, we did away with gasoline altogether and became sort of a multipurpose hardware store. Conoco's retail strategy today owes part of its growth and change to Western's experience."

McCollum also intensified Conoco's marketing strength in California by buying Douglas Oil in 1961. Douglas, which operated three small refineries and sold refined products through some 300 retail stations, was in the middle range in terms of price and service. Altogether, the four acquisitions gave Conoco a broad spectrum of retail strategies in its various markets. "When it comes to selling gas," McCollum often said, "we'll be Neiman-Marcus or we'll be Woolworth — whatever the customer wants."

Conoco's spending wasn't confined to the United States. McCollum simultaneously engineered a similar series of downstream acquisitions in Europe, putting together an acquisition team composed of seasoned executives in Europe whom he had hired away from Caltex, a competitor with substantial European interests. "It seemed like we were handling an acquisition a week back then," a beleaguered executive from the company's legal department later said. The Libyan crude was flowing, but Conoco was unable to exploit its potential. U.S. import restrictions and Conoco's still-modest domestic refinery capacity required it to sell much of the crude locally to independent buyers or international oil companies that, unlike Conoco, already had a marketing presence in Europe. McCollum realized the only way to garner a worthwhile profit from the company's upstream investment in Libya was to go downstream as well. With a series of startlingly swift acquisitions, he built Conoco overseas into an integrated oil company.

In 1960, Conoco acquired Sopi, a petroleum marketer with more than 400 retail stations in West Germany and Austria. With no time to build its own refinery in Europe, Conoco negotiated a 20 percent interest in a recently completed refinery in Karlsruhe, West Germany, and made arrangements with refineries in Sicily and mainland Italy to process Libyan crude for sale in Italy and for export to northern Europe. Conoco also acquired a small interest in a major European pipeline, Trans-Alpine, and developed plans for another pipeline to be laid from the port of Genoa to Milan, where an oil distribution and storage terminal later was erected on a 200-acre site.

Arrangements also were made at the time to acquire Seca, a strong independent marketer with about 100 retail stations in Belgium and Luxembourg.

Then, in 1961, Conoco purchased Jet Petroleum Limited, an independent distributor of petroleum products in the United Kingdom with more than 400 retail outlets. Jet was started in 1954 by an entrepreneurial coal merchant from Yorkshire who saw an opportunity to sell low-cost gasoline, according to Roy Roley, Conoco's brand commercial manager in Warwick, England, current home of the company's downstream U.K. headquarters.

"By sheer fluke," says Roley, "the license plates on the tankers said 'JET,' so that's what he called the product. He cut sixpence off the cost of a typical liter of gas, and, like its name, it took off."

By the time the Es Sider pipeline from Dahra to the Libyan coast was completed in 1962, Conoco's downstream operation was up and running. By 1968, Conoco's one-third share of the one billion barrels of Libyan crude flowed from Es Sider to refineries in Italy and Germany and, ultimately, into the waiting cars of consumers at Conoco's 900 Jet, Sopi and Seca stations. To serve these expanding international markets, Conoco purchased four tankers: the *Dubai*, *Seca*, *Sopi* and *Arrow*. Two of the vessels delivered oil to European depots, as did the *Libya*, which the company purchased later in the decade. The other ships worked in the "clean trade," delivering refined petroleum products to markets. From this solid foundation, the company would continually enlarge its presence throughout Europe and, ultimately, into markets all over the world.

Conoco officers meet with the ruler of Dubai, above. In 1963, Conoco formed the Dubai Petroleum Company, which acquired a 914,000-acre concession covering segments of the mainland and territorial waters of Dubai. The company undertook exploratory drilling, and a major discovery was made in the Fateh field, in waters about 55 miles (88 kilometers) off the coast, opposite. The emirate, then a small, relatively unknown gulf port, issued a commemorative stamp, left, featuring the Fateh field.

By 1965, Conoco was pumping more crude outside the United States than domestically, boosting its earnings over $100 million for the first time in its history. Within a short time, more than Libyan crude would course through Conoco's overseas network. In 1963, Conoco became one of the first independent oil companies to break into one of the coveted areas surrounding the Arabian Gulf. The company acquired a 914,000-acre onshore concession from the ruler of Dubai on the Trucial coast.

This first onshore contract enabled Conoco to obtain from a British and French company 50 percent of the exploratory and production rights covering all of Dubai's offshore area. Conoco formed the Dubai Petroleum Company, which became the operator for both the onshore and offshore areas. To diminish risk, Conoco conveyed a 10 percent interest to another American company, and subsequently, a Spanish company acquired another 5 percent interest.

At the time, the Dubai emirate was a small, relatively unknown gulf port. "There wasn't much infrastructure," recalls Conoco retiree Stan Conly, who later became general manager of employee relations. "The runway was just long enough for jets to land, and there was only one paved road — the rest were hard clay. The water was a bit brackish but safe, and my wife, Susan, and I stayed in a cinder block house with central electricity. Fortunately, air conditioning had just been installed.

It was hotter than blazes in the summer!"

Nevertheless, seismic operations were undertaken, and an exploratory well was completed in early 1964. "We drilled four holes and spent quite a bit of money," says retiree Harry Sager, who later became executive vice president of exploration and production. "We were just about to call it quits, when we decided to drill one more hole in 1966. It was a great producer." The discovery was made in the Fateh field, in waters about 55 miles (89 kilometers) off Dubai. A second discovery in the southwest Fateh field in 1970 would further boost the total output of Dubai crude.

With the purchase of the *Italia* during the mid-1960s, Conoco began transporting Middle Eastern oil to the United States (and U.S. coal to Europe).

As if to underscore the company's new role as a fledgling international player, McCollum moved Conoco's headquarters from Houston to the nexus of global commerce — New York City. "The hunt for oil would take place primarily outside the United States, and McCollum wanted proximity to all that New York offered. Having come from Standard Oil of New Jersey (now known as Exxon), he felt this was the way to demonstrate Conoco had 'arrived' as an international company," Dino Nicandros says. Dick Hittle agrees: "New York City at the time was the absolute center of the oil industry. Most of the majors were there, including Exxon, Mobil and Texaco. McCollum knew quite simply that Conoco just had to be there."

Conoco took up residence at 30 Rockefeller Center, within view of the famed bronze statue of Prometheus. In Greek, the name means "foresight." It was a fitting metaphor for the mid-size oil company that had leveraged the opportunities of global enterprise.

In the search for oil, every company needs a guide. John G. McLean was Conoco's chief international business developer, appointed by McCollum to direct its international program in 1964. McLean, a Harvard professor and later chief executive officer of Conoco, favored an academic approach to exploration. In 1966, he formed Conoco's Advance Geological Group and set it up near Princeton University. The group became a geological divining rod guiding Conoco to promising oil sites in Indonesia and in its recently acquired acreage in the North Sea.

"We were the oil detectives," recalls John W. Strickland, Conoco's retired chief geologist and the first manager of the group. "We studied the sedimentary basins of the world to determine which areas had the most potential for petroleum exploration. Princeton's geological library provided invaluable resources as far as where these basins were located."

For competitive reasons, the company followed the practice of using code names in all memoranda when discussing favorable oil properties or its scientific research. Research was the key to unlocking the oil reserves of tomorrow, and neither the "oil detectives" nor Conoco's R&D department wanted any leaks.

In fact, after its reinstitution by McCollum, the research department helped Conoco achieve a number of industry firsts, including the first delayed coking process (at the Ponca City refinery), the first quadruple-completion gas well (by CATC), the first automated offshore production system controlled from shore (in the Gulf of Mexico), the first underwater pipeline connection (also in the Gulf), the first synthetic cold-weather motor oil (for the U.S. government) and the first production platform

JON-AL DUPLANTIER

After three months in the desert, Jon-Al Duplantier misses the relatively balmy evenings in his Cajun hometown of Grambling, Louisiana. But despite the extreme weather in the United Arab Emirates, Duplantier — Conoco's vice president and general counsel in Dubai — relishes his first overseas assignment. "Dubai is very cosmopolitan and has invited the citizens of many lands to live and work in the country," says Duplantier. "I like the diversity. And I like the feeling of community." Duplantier was transferred to Conoco from DuPont's legal department in 1995 and helped prepare the initial registration statement for Conoco's independence. "The closer we got to the IPO, the more frantic the work became. We'd do our jobs during the day and spend all night at the printing company." When the IPO finally reached fruition, "the New York Times was writing about it," says Duplantier, "and here we were still doing the final touches on it. I remember feeling the biggest adrenaline rush of my life."

GOING GLOBAL 119

Pictured are Conoco laboratory technicians in Lake Charles, Louisiana, creating synthetic products from crude oil. Several petroleum-based alcohols, including Alfol and Nalkylene Alkylate, were manufactured by the company in the 1960s and sold to the detergent industry. Together these products helped Conoco capture 25 percent of the synthetic detergent materials market in 1964.

in more than 100 feet (30.5 meters) of water.

It was McCollum, too, who brought petrochemicals to the forefront of Conoco's research efforts in the late 1950s and 1960s, as company researchers experimented with other synthetic products derived from crude oil. None were more promising than the petroleum-based alcohols developed by Conoco. These chemical intermediates were used in the manufacture of detergents, plasticizers and other products. Conoco was the first company to develop a biodegradable intermediate for detergents that would decompose in sewage treatment plants; it was called Nalkylene Alkylate. The environmentally sound intermediate was produced at its refinery in Baltimore. At the Lake Charles refinery, Conoco produced another biodegradable intermediate, developed in 1960 and trademarked as Alfol.

These products helped Conoco capture 25 percent of the synthetic detergent materials market by 1964, making it the leading supplier in the United States. "We called it the suds business," said Frank Diassi, a Conoco salesman in Chicago in the 1960s. "Our intermediates were used in most of the top products, including Tide, Palmolive and Ajax." In 1962, the company established a joint venture with a German firm to construct a plant to manufacture Alfol for sale to European detergent makers. In the early 1970s, Conoco sent Don Butter and Joe Kotarski to Moscow to license the Alfol technology to the U.S.S.R. "at a time when it was not necessarily popular to do business with the U.S.S.R.," recalls Butter. The licensing agreement ultimately netted the company $12 million. By 1975, Conoco Chemicals, a subsidiary, was the largest manufacturer of detergent intermediates in the world.

Other petrochemical ventures also proved encouraging. The Anotrol process, a method for inhibiting the deterioration of metal in contact with corrosive acids, was invented by Conoco researchers in 1960, and its components were produced at the Baltimore refinery. The company also was active in plastics and plant foods. In 1963, it purchased a controlling interest in Carlon Products Corp., the largest U.S. manufacturer of lightweight plastic pipes. That year, Conoco also purchased American Agricultural Chemical Company, a fertilizer manufacturer. McCollum saw a good fit, as hydrogen sulfide, a by-product from all of Conoco's petroleum refineries, is easily processed into sulfur — an important ingredient in fertilizer. Under McCollum's leadership, Conoco had started down the path to becoming a leading supplier of specialty chemicals.

A few months after transplanting the company's headquarters to New York in 1964, Conoco's board of directors shuffled its corporate management. Charles Perlitz, Conoco's board chairman, had passed away in August, leaving a vacancy. Asked to fill the void, McCollum, who retained his CEO title, accepted. The presidency was passed to Andrew W. Tarkington, a tall, cool-eyed Texas banker who came to Conoco in 1948 as its treasurer. Tarkington would become the company's CEO in January 1967. Until then, McCollum was firmly in charge of company operations.

A pressing issue of the day was the desire among oil companies to become more broadly diversified energy suppliers, particularly to the nation's many electric power utilities. Several oil companies, including Texaco and Gulf, already were producing coal and uranium, two raw materials used for electric power generation. McCollum didn't want Conoco to be left behind. "If you don't change with the conditions, you go down the drain," he once said.

On September 15, 1966, Conoco officially gained a stake in the coal industry, buying the Consolidation Coal Company, the second-largest U.S. producer of bituminous coal. Consol, as it was known in the industry, was even older than Conoco, dating its origins to 1860. The company owned 36 bituminous and lignite coal mines, which brought in $307 million in gross revenues and $33 million in net income in 1967 — about a third of Conoco's revenues and income at the time. Half of Consol's production went to electric utilities, with the steel industry ranking as its second-largest customer. The *New York Times* hailed the complex $620 million merger, which again involved highly innovative off-balance-sheet debt financing, as "historic."

Hottest Brand Going ...CONOCO!

This award-winning advertising campaign from the late 1950s and early 1960s has branded itself into the public memory. The ads drew from the company's pioneering spirit and Western roots, fostering an enduring image of ruggedness and independence.

Coal research was the top priority of the combined company, as company scientists sought ways to convert coal into other energy forms, especially gasoline and high-BTU gas. Consol's experimental coal-to-liquids plant in Cresap, West Virginia, perfected a process whereby coal was ground and partially dissolved in a special extractor, and the extract then was submitted to catalytic hydrocracking under pressure. The resulting liquid was enough like crude petroleum that high-octane gasoline could be produced from it using standard refining methods.

In 1967, Consol joined with the U.S. government at a pilot plant in South Dakota to research the efficacy of converting coal into a synthetic alternative to natural gas. While optimism was high at the time for all these coal-to-gas plans, they were never successfully realized on a commercial scale. "It's a long road from the laboratory to commercial production," company historians acknowledged in 1975.

Coal was not the only mineral Conoco extracted from the earth. The company also was active in mining both uranium and copper. "We felt we had the geologic talent to get into these minerals, which at the time were commanding fairly high prices," says Aivars "Ike" Krasts, who retired from Conoco in 1996 as its vice president of corporate stragegy. "We had the manpower, talents and equipment in every phase of the action, from exploration to research, technical development and even financing."

In September 1967, McCollum formed Conoco's Uranium Exploration Group. Over the next eight years, the group would account for 10 percent of all uranium drilling in the United States. In 1968, the group discovered significant uranium ore deposits in South Texas that led to the startup of a mill at the site. Completed in 1972, the mill supplied raw material for the manufacturing of nuclear fuel rods. In 1971, the company also reported success in its copper operation — the discovery of a large copper deposit near Florence, Arizona.

When he retired as the company's chairman in 1972 at the age of 70, L. F. McCollum was honored as one of the outstanding oil visionaries of the century. He had transformed Conoco from a regional oil company with total assets of $200 million and net income of about $30 million in 1947 into an integrated, worldwide enterprise with more than $2.3 billion in assets and $136 million in net income in 1967, the year he stepped down as CEO. Under his leadership, the company had diversified into coal, chemicals, plastics, fertilizers and minerals. Key corporate activities such as R&D, finance and planning had been greatly expanded.

When he passed on his chairman's title, Conoco was listed in the top 25 of the Forbes 500.

"He was one of the towering figures of the oil industry," Dino Nicandros says of his mentor, who died at the age of 91 in 1993. "He had the vision to get into things when others thought it unwise, from Libya to Consol, and it turned out they were the right things to do. Above all, McCollum was a builder. He took this middling, regional oil company and made it into a truly competitive player. And he did it with a rare combination of guts and grace."

McCollum once said, "I get as much satisfaction from seeing a man succeed in a new job as from discovering a promising oil field." Reminded of the quote, Nicandros pauses in thought. "You know, it takes courage and confidence to delegate responsibility," he says. "McCollum truly enjoyed and thrived on collaboration. He surrounded himself with talented individuals, and he nurtured them to think creatively. Part of his legacy is that the people he picked to join his journey became its next generation of stalwart leaders."

As a testament to McCollum's historic impact on the company, Conoco assembled a display case honoring him at its headquarters in Houston. Among the many artifacts presented is a pair of Mr. Mac's world-weary cowboy boots — heels and soles that touched the soil of Libya, Dubai and other faraway places. The tales they could tell.

L. F. McCollum transformed Conoco from a regional oil company into an integrated, worldwide enterprise. To honor Mr. Mac after his retirement as the company's chairman in 1972, the company erected a display case containing artifacts from his years at Conoco, including his much-traveled cowboy boots. It was a symbolic gesture that conveyed both gratitude for the grand strides he helped the company take and recognition that Mr. Mac's shoes could never be filled.

Crisis and Competence

When the Arab oil embargo in the winter of 1973 caused an international energy crisis — made painfully evident in the United States by the interminable lines at gas stations — Conoco fortunately had John Godfrey McLean to guide it.

Named Conoco's 10th president in August 1969, McLean, a full professor from Harvard Business School, was the chairman of the government committee that originally predicted the crisis. Known for his brilliant mind and extraordinary capacity for analysis and concentration, McLean had come to Conoco as a Harvard consultant in 1949. Five years later, L. F. McCollum offered him a full-time job. "I knew he was a workhorse, and I wanted him as my assistant," McCollum said. "He took a leave of absence from Harvard, but he never went back. The oil business got to him."

When Conoco employees talk about McLean, they speak admiringly about his intellect and vitality. Then, inevitably, one of them will tell the story about how McLean, the son of a Presbyterian minister, was in Germany on a business trip, and a group

of executives invited him to join them at a nightclub. While the other executives were socializing, McLean spent the evening hunched over the table, sipping a glass of bitter lemon and making notes for the next morning's meeting. "He genuinely loved his work, and I don't think anybody ever worked harder," said Samuel Schwartz, a student of McLean's at Harvard who worked closely with him as Conoco's vice president and general manager of coordinating and planning.

McLean took the reins from Andrew Tarkington, who had succeeded McCollum as president in 1964 and CEO in January 1967. Tarkington's background in banking and finance helped him reshape the company into a more efficient business machine. He consolidated operations, closed several regional offices, established new priorities for existing projects and introduced new financial techniques for evaluating business risks.

These were prudent moves; dependence on foreign oil was growing, and credit was no longer so readily available as it had been in the 1950s. Serious problems also threatened overseas. In June 1967, war erupted in the Middle East between several Arab nations and Israel. Tarkington's response was to chart a more deliberate, conservative course. "Andy felt we couldn't continue expanding at the rate we'd been going," says Dino Nicandros. "Some projects were put on the back burner, and several were canceled or rejected."

Tarkington continued to push for innovative and cost-saving solutions, however. One such solution was championed by Conoco engineer, L. B. "Buck" Curtis. In August 1969, a 28-million-pound underwater storage tank shaped like an inverted champagne glass was towed 58 miles (93 kilometers) from a construction site in Dubai to the Fateh field in the Arabian Gulf. Once there, the tank was gently submerged until its bottom rested 150 feet (46 meters) below the surface on the seafloor and anchored firmly in place.

As Conoco entered the 1970s, more than just its familiar logo would topple (changed from the old Marland red triangle to today's familiar red and white capsule). An oil embargo by the Organization of Petroleum Exporting Countries (OPEC) in the Middle East fostered public animosity toward the U.S. oil industry, which was blamed for the resulting shortage and the interminable lines at gas stations. The crisis prompted the U.S. government to intervene with crippling price controls and threats of further regulation. Conoco employees across the globe fought the tide of negative opinion with facts and figures proving the industry was a scapegoat for OPEC's actions.

The Khazzan underwater storage tank, pictured here ready to be submerged off the coast of Dubai, is a Conoco technological marvel. Built as an alternative to the expensive pipelines and onshore storage tanks traditionally associated with offshore oil production, Khazzan saved the company millions of dollars. The technology, developed by L. B. "Buck" Curtis is based on the principle that oil floats on water. Curtis later pioneered an innovative platform that enabled more economical deepwater exploration and production. The basis of the new technology was the tension leg, demonstrated by Curtis below.

128   125 YEARS OF ENERGY

Dubbed Khazzan, the first-of-a-kind tank was the company's alternative to building the expensive pipelines, onshore storage tanks and docking facilities traditionally associated with offshore oil production. Altogether, it shaved millions of dollars off the cost of transporting and storing Conoco's Dubai crude.

"We utilized the principle that oil floats on water," says Curtis. "As oil is pumped into the storage facility, the water in the dome is forced out at the open base. The reverse process then takes place when oil is pumped out of the tank into the holds of tankers."

First, Curtis had to sell Conoco's board of directors on his idea. Wayne Glenn, then president of Conoco's Western Hemisphere petroleum division, remembers the novel way Curtis demonstrated the concept: "Buck brought in this tank of water — actually a kid's pool from Macy's — and he floated out this little toy that looked like the bottom of a plumber's plunger and flipped it over. His eyeglasses hung down on the tip of his nose, and he looked so darn serious. And, doggone it, it worked just like he said."

The board gave the go-ahead to the project to develop the Fateh field, which ultimately cost $62 million. Two more underwater tanks were subsequently constructed and submerged in similar fashion.

Nevertheless, beset by economic and political events beyond his control, Tarkington had to make some difficult decisions that didn't go over well among managers used to McCollum's more giving ways. McCollum was a very tough act to follow, and though Tarkington was CEO, McCollum, still chairman, was always looking over his shoulder. It eventually got to be too much. In August 1969, after 21 years with the company, Tarkington announced his retirement from operations management, but he retained his board of directors membership.

John McLean was running Conoco's international operations in New York when he was called in to replace Tarkington as president and CEO. In planning a strategic course for the company, McLean focused on the mix of energy-producing resources that he believed would be most meaningful to its future. "We intend to develop strong positions in oil, gas, coal, uranium and copper and to keep our activities closely confined to these areas," McLean said upon taking office.

He emphasized that exploration would continue to be Conoco's engine of growth. One area that appeared to offer promise was the North Sea, the arm of the Atlantic between the United Kingdom and the European mainland. The company had acquired significant exploration acreage in the Southern Basin of the North Sea in the first round of U.K. licensing in 1964. Four years later, the Southern Basin acreage produced the find that would be the foundation of the company's North Sea producing business.

As president of Conoco, Andy Tarkington faced a changing climate in international oil exploration, including warring nations and tightening credit. His background in banking was the perfect foundation for his leadership role. He led Conoco through five years of belt tightening, setting new standards for evaluating risk, strengthening corporate services and consolidating assets wherever possible. He applauded cost-saving innovations such as the Khazzan and alternative fuels research, such as coal-to-gasoline, that could lessen U.S. dependency on imported oil. Tarkington retired in 1969.

When Conoco sought a location for a European refinery to process its low-sulfur Libyan crude, it chose a site on the Humber River, located on England's North Sea coast. The Humber refinery, completed in 1969, set a new quality standard for European refineries and featured units to produce chemical feedstock and petroleum coke. Today, Humber is ranked as the top European refinery by Wood Mackenzie and is one of the top suppliers of coke, below, to the steel and aluminum industries.

McLean also presided over the formation of a Conoco-National Coal Board (NCB) of Britain consortium that discovered 3 trillion cubic feet of natural gas in several fields in the Viking area off the Norfolk coast. Four years later, the field was delivering gas to the fledgling U.K. gas distribution system. Natural gas was transported by a transmission pipeline to a new Conoco-NCB Viking gas terminal at Theddlethorpe on the Lincolnshire coast, southeast of the Humber River estuary. The Viking terminal fed British Gas Council distribution lines serving London and other markets with clean natural gas.

Viking was Conoco's first major upstream development outside Libya, Dubai and North America. As it turned out, the Viking discovery began a chain of events that would silence critics of the siting of Conoco's first wholly owned refining and petrochemical complex outside the United States — the Humber refinery, also located in Lincolnshire.

In the mid-1960s, the company had been looking for a place to build a refinery to process its low-sulfur Libyan crude. England was a natural, given the company's significant downstream presence there. A site was selected in 1966, and construction began in earnest. Three years later, thanks in large part to the tireless efforts of future Executive Vice President Bob Turvey — the "father of Humber" — the refinery was ready.

Until that time, European refineries basically had produced a lot of heavy fuel oil for industry, as well as gasoline and middle distillates. Humber would change all that, its configuration setting a new standard for European refineries. Turvey led the Humber team as it constructed a U.S.-style refinery configured to produce almost no heavy fuel oil and almost all gasoline, middle distillates and petroleum coke. This configuration, and an original capacity of about 80,000 barrels a day, quickly made Humber the most talked-about and profitable refinery in Europe.

Conoco increased refined petroleum product deliveries in the late 1960s, purchasing two vessels — the *Jet* and the *Humber* — to shuttle products from the Humber refinery to European ports. Two tankers built in Spain — the *Britannia* and the *España* — were purchased to transport the Libyan crude to the refinery as it expanded in the 1970s.

Serendipitously, Humber just happened to be sitting smack dab on the North Sea on England's northeastern coast. "When we built Humber, everyone figured we'd built it on the wrong side of the country," says Gary Edwards, Conoco executive vice president for downstream. "Gulf, Texaco and other oil companies had selected the deepwater ports off the southern coast of Wales for their U.K. refineries. We were offered some tax benefits if we'd choose the other side — a more depressed region of the country — and we took them. Next thing you know, the North Sea is a major oil- and gas-producing region. I'd like to say we knew this would happen, but I'd be lying. It's just one of those things where blind luck later looks like genius."

The company suddenly had a foothold in a region of the world that offered great promise and, unlike the Middle East, few political concerns. As tensions simmered in the Arab world, Conoco's international exploration and production department would negotiate a series of deals that would position Conoco as a major player in the North Sea. The crude oil that flowed from these investments kept the company afloat in what would turn out to be an extremely turbulent decade.

The North Sea's southern waters were rich with natural gas deposits, but essentially no oil. Geologic studies indicated significant oil-bearing strata were more likely to occur in the East Shetlands Basin to the north. The London-based exploration team, now headed by Lloyd Ryman, recommended an aggressive acquisitions effort. McLean agreed and gave his former London colleagues the mandate — go north.

"John told us we could have whatever we needed for an aggressive exploration program in the North Sea," said Mike Morris, who in the early 1970s was an executive vice president in Conoco's Eastern Hemisphere petroleum division. "It added up to a huge capital commitment. We hit a dry spell in the early 1970s when we weren't getting much for our drilling

A Conoco-National Coal Board (NCB) of Britain consortium discovered the Viking field in the southern waters of the North Sea off the Norfolk coast. Viking, a 3-trillion-cubic-foot natural gas field, was Conoco's first major upstream development outside Libya, Dubai and North America, and Conoco's relationship with the NCB would prove critical to future deals in the North Sea.

money, but John never let himself be discouraged."

In 1970, the international group negotiated a landmark deal in the company's history, a deal that positioned Conoco to own a major equity interest in the largest oil field ever found in the North Sea — the Statfjord field. "Our London team worked out a farm-in with Gulf Oil whereby they conveyed to Conoco a one-third interest in their important northern U.K. exploration holdings and agreed that Conoco would become operator," recalls Ryman.

It was a heck of a deal. Gulf was in partnership with Britain's National Coal Board on the northern U.K. blocks, and Conoco had a good relationship with NCB from its Viking gas venture in the south. Gulf had drilled 14 dry holes and was on its 15th well when the deal was made at the handshake level. While the deal was being finalized, the 15th well encountered oil. To Gulf's credit, they honored the handshake.

Before Statfjord was discovered in 1974, Conoco had drilled a well on the U.K. side of the international boundary. "As Norway called for submission of bids for the block that held the heart of the seismic prospect, intense competition favored the large oil companies," says Dennis Gregg, vice president of exploration and production international. "But what Conoco lacked in size it made up for in successful offshore experience in the Gulf of Mexico, Dubai and the southern North Sea. And it had already contributed to the exploration effort with the dry hole on the U.K. side."

Conoco also already had a small but important position in Norway acquired through a farm-in to Texaco acreage, but the acreage holding the Statfjord prospect was a prize of much larger potential. Marvin Lesser, Conoco's first expatriate representative in Norway, and Ryman convinced the Norwegians that the company was just as capable as Esso, Mobil and Shell, despite its smaller size. Norway awarded Conoco a 10 percent equity in the prize blocks, the same as Esso and Shell. Mobil, then the project operator, received a 15 percent interest, and a group including Amoco and Saga received 5 percent. Norway's Statoil, the state-owned oil company, assumed a 50 percent share and later became operator of the field. "Conoco later drilled a successful well and proved that 15 percent of the field was on the U.K. side," Gregg says.

With its combined U.K. and Norwegian interests, Conoco had the largest share of any private company in the field. "Had we not negotiated the Gulf and Texaco transactions, underscoring our commitment to the North Sea, or demonstrated our international operating strengths, Conoco would not have had an equity in Statfjord. Statfjord and our other North Sea interests constitute one of the most valuable assets owned by private oil companies," says Dick Hittle.

Conoco had become a leader in the North Sea and, over the next 20 years, would build substantially

Conoco's growing reputation as an international, integrated oil company, coupled with superb negotiating, earned it a major share in the largest oil field ever found in the North Sea — the Statfjord field. Participants in the Statfjord field were a veritable "Who's Who" of oil companies, as seen in the publicity material to the left.

CRISIS AND COMPETENCE   133

In 1972, Conoco moved its headquarters from New York's Rockefeller Center to the Stamford, Connecticut, building pictured here. The move was prompted by a steep rise in office costs and the desire of employees to reduce their commute. From this gleaming, glass-enclosed facility, the company would plot its course through much of the difficult decade.

By the time Conoco's headquarters moved to Stamford, Connecticut, in 1972, following a sharp rise in New York City office costs, the company was ranked number 23 in the Forbes 500 largest companies in the United States. L. F. McCollum retired that year, and on its holdings and its reputation. Though still small compared with the majors, Conoco was increasingly perceived as a player with clout.

McLean added the chairman's post to his duties.

Significant changes were taking place in the energy business. Supplies were becoming tighter and the risk of a serious shortage in the near term was increasing, as was the likelihood of not being able to satisfy future demand. Always the professor, McLean recognized that the oil industry needed to articulate these changes to the government and the public. So in 1972, he accepted the chairmanship of the Committee on the U.S. Energy Outlook for the National Petroleum Council (NPC) and supervised the compilation of the committee's report. In bleak terms, it predicted an imminent major oil crisis.

Conoco was among the first companies to alert the public to the impending energy crisis. In 1972, the company published a message titled, "Energy & America," that appeared in many leading U.S. publications. The directive reiterated the dire warnings of the NPC's report and attracted thousands of letters and requests for information. Before these could be disseminated, however, the committee's prediction became reality. An embargo during the winter of 1973 by the Organization of Petroleum Exporting Countries (OPEC), which represented major oil exporters like Saudi Arabia, Libya, Iran and Iraq, caused a fuel shortage so severe that many factories in the American heartland closed and gasoline prices skyrocketed.

Conoco's report suddenly had the immediacy of the morning newspaper and the evening news. It answered the three basic questions that were on everyone's minds: How much energy will be needed in the years ahead; where will this energy be obtained; and what changes in government policies will be required to get through the crisis?

The study concluded that the U.S. had enough domestic fuel resources to satisfy energy demands, but developing these resources would require higher prices and farsighted national energy policies. Increased reliance on imports to meet these energy requirements would be futile, given the uncertainties regarding both availability and price. Restrictions on energy demand would be equally undesirable, altering lifestyles and adversely affecting economic growth, employment and freedom of consumer choice. Even more efficient use of energy could not quell the crisis.

Instead, the report advocated a comprehensive national energy policy, a realistic approach to environmental issues, accelerated leasing of public lands for exploration and increased development of synthetic fuel alternatives. Conoco already was seeking more efficient alternative energy production through its coal gasification and other projects. These initiatives were stepped up. The company also announced the most extensive foreign exploration program in its history in 1974, albeit in countries generally outside the Middle East. Exploration was under way in Chad, Central African Republic, Madagascar and Qatar.

John Godfrey McLean, a former Harvard professor reputed to have an extraordinarily high IQ, was named Conoco's 10th president in August 1969. McLean was hired briefly by Conoco as a consultant in 1949 and was later offered a full-time job by then-CEO L. F. McCollum to help develop the company's international business. McLean recognized early that an imbalance in oil supply and demand was inevitable and that U.S. reserves could not compensate. His prediction of an imminent, major oil crisis was right on target. Unfortunately, illness cut short his tenure.

CRISIS AND COMPETENCE 135

In the United States, Conoco and other petroleum companies were limited in their ability to explore for oil. They could not drill offshore for oil and gas, build refineries, lay pipelines or plan deepwater ports in the United States without encountering extensive regulatory measures. Coal and uranium mining faced similar bureaucratic obstacles. Thus, at a time when its ability to import foreign oil was hampered, the industry was unable to effectively increase its domestic production.

The dire straits that John McLean had prophesied were the talk of the nation. But the man who had predicted the crisis would not live to see the company through the end of the turmoil. In May 1974, at the age of 56, McLean died after a brief illness. Conoco retiree Tom Sigler offered this perspective on McLean and his two predecessors: "Mr. Mac took the company and made it big and strong. He was a builder. Andy Tarkington was a transition man. He tightened up the operation and established us securely as a multinational energy company.

"John McLean continued building, but he did much more. He brought a kind of calmness to the company. He encouraged people to say what they had in mind. And because of his great intelligence, he brought an aura of intellectuality to the company. He made you proud to be a part of Conoco."

McLean's five-year term as president is keenly felt a quarter century later. In his first few weeks in office, he made a pledge to establish a sound environmental policy for the company, a standard of excellence that persists. He took positive action to install pollution-abatement equipment, to comply with environmental legislation and to establish an environmental council of specialists from different disciplines and units of the company. "No single goal set for the decade of the 1970s ranks higher than winning the war against pollution," McLean once said.

He also put a premium on research and education, not surprising given his master's and doctoral degrees from Harvard. He also continued the company's tradition of technological firsts. During his tenure, Conoco patented a technique for gravel-packing wells, was the first to use a new type of semisubmersible rig in the North Sea capable of drilling in water depths up to 1,000 feet (305 meters) and developed the first polymeric drag reducer for pipelines. It also was the first to use a computer to evaluate seismic data, developed Japan's first delayed coking plant and, a few days before McLean's death, reported the first successful coal gasification, a testament to the ingenuity and determination of the company's synthetic fuel research.

Although he did not live to see the company through the rest of the tumultuous decade, a period in which its mettle was tested time and again, John McLean had prepared Conoco well, setting in motion the policies and programs it needed to weather the trials ahead.

Environmental concerns about the effects of leaded gasoline led the U.S. Environmental Protection Agency to mandate the development of a lead-free gasoline. Conoco's refineries were modified in the mid-1970s to comply with the EPA's regulations. A sound environmental policy was one of McLean's highest achievements with the company.

CONOCO ON THE HUNT

McLean anticipated world energy shortages and took action. This prediction was fulfilled by the 1974 OPEC embargo, which underscored the critical need for a broader and more extensive exploration program. That year, Conoco began the most expansive program of oil exploration in its history. From Qatar and Indonesia to the Gulf of Mexico and the North Sea, Conoco's oil hunters would discover new supplies of crude and natural gas to meet accelerating demands. Conoco gained a reputation for innovative and cost-effective exploration and project management. Here Conoco's exploration teams are seen in the various climes of (clockwise from top) Louisiana, the Southwestern U.S. and Africa.

Howard W. Blauvelt began his career as Conoco's assistant controller in Ponca City, Oklahoma. Praised for bringing financial discipline to Conoco but criticized for his unsympathetic handling of executives, Blauvelt coolly guided the company through the Middle Eastern oil crisis. He is credited with positioning Conoco for the future by spending more than $5 billion in capital investments and exploration expenses on the company's burgeoning oil, coal and chemical operations. Although he disdained being referred to as a "former accountant," Blauvelt's strict financial controls within the company and his willingness to invest outside the company exemplified the business acumen associated with the best of the "Big Eight" oil companies at the time.

Howard W. Blauvelt, a 22-year Conoco veteran, was elected to succeed McLean as chairman and chief executive officer in May 1974. New York-born and Yale-educated, Blauvelt had started his career as an assistant controller in Ponca City and served stints with Hudson's Bay in Canada, Consol and Conoco Chemicals. He brought a wide range of experience to the job.

Blauvelt wasn't afraid to shake things up. In 1975, he initiated a sweeping reorganization of the company along functional rather than geographic lines. Anticipating a more global industry, he sought to create a stronger central operation, further integrating Conoco's Houston-based Western Hemisphere petroleum operations into the growing international company. This group, headquartered at Conoco Tower in Greenway Plaza, customarily produced two-thirds or more of the company's consolidated earnings and had become near autonomous. Blauvelt wanted more control vested in Conoco's headquarters group in Stamford, and the board agreed. Though some executives resisted the reorganization, Blauvelt was less concerned with soothing tempers than dealing with the effects of the Arab oil embargo that began in the winter of 1974.

OPEC curtailed the production of oil from member countries, creating a supply-demand imbalance that quadrupled the per-barrel price of oil sold to the international oil companies. The industry had no recourse but to increase prices at the pump. The public, however, didn't accept this cause-and-effect scenario. A mindset developed that the crisis did not exist and was, in fact, a conspiracy by the big oil companies to cooperate rather than compete with one another — to maximize profits and squeeze out independent, nonintegrated oil companies. Abetting this perception was the huge increase in revenues the oil companies reported while the embargo was in effect.

The industry responded forcefully to these charges in print and before Congress, but the negative conclusions coalesced into a countrywide consensus. People needed to point fingers, and the oil industry became the scapegoat for OPEC's action. Almost overnight, "the oil companies were now among the most unpopular institutions in America," wrote author Daniel Yergin in *The Prize*, a history of the oil industry.

Many current Conoco employees recall the emotion and outrage. "In the early 1970s, I was responsible for selling Conoco's production of propane, which is used as a home heating fuel," recalls Archie Dunham, natural gas products manager at the time and later chairman, CEO and president of Conoco. "As shortages developed, concerned constituents called their local legislators, who then called the company and demanded action. We did what we could, given the circumstances." Conoco and other

Conoco Tower in Houston's Greenway Plaza office park, pictured at right, was Conoco's petroleum operations headquarters during the years 1974 through 1984. This tumultuous period saw the OPEC embargo, nationalization of many of Conoco's properties, gasoline shortages and price controls in the U.S., the Iranian crisis, the merger with DuPont, and the first price slump of the 1980s. Despite these difficulties, Conoco continued to expand, and in 1984, the company moved its headquarters from Greenway Plaza to Conoco Center at Woodcreek, a much larger facility able to house increased staff requirements.

The national gasoline shortages that followed the Arab oil embargo forced retailers to ration their supplies and motorists to wait in long lines at the pump. The shortages made exploration an urgent priority for U.S.-based oil companies.

oil companies urged industrial users of natural gas to convert to other fuels, such as heavy fuel oil or coal, to ease the gas shortage. Conoco set the example by initiating conversion projects for two of its own refineries.

Individual salesmen and marketing managers worked to supply fuel for emergency requests. In one day in 1974, for example, Max Punches, a Conoco division marketing manager, located fuel in another city to fill the tanks of Billings, Montana's, emergency vehicles, helped the governor of Idaho respond to questions from businesses and ranchers in his state and found additional fuel to avert the closing of a sawmill in a small Montana town. In addition to these emergency measures, Conoco urged all employees to speak out to educate consumers about the fuel supply situation, stressing conservation and explaining the factors that were causing the shortage.

Nevertheless, legislators sought increased regulation of the industry. Says Dunham, "I remember going to Washington in the early 1970s and meeting with government regulators. They were extremely adversarial and, frankly, knew very little about the industry. Yet they were given responsibility for preparing a set of regulations determining how we should distribute our product."

The regulations were stringent and burdensome. U.S. President Richard Nixon had imposed price controls on all sectors of the U.S. economy in 1971 in an effort to reduce inflation. Although most such controls expired in 1974, those on oil remained in place. In effect, the regulation prevented companies from importing oil at prices above government-mandated levels. For example, when Conoco tried in May 1973 to import 26,000 barrels a day of gasoline and fuel oil from Europe to meet its customers' needs, the government's Cost of Living Council denied the application because the foreign oil averaged about one cent a gallon over the price ceilings then in place.

"Our customers are being denied some 46 million gallons of products a month," said Howard Hardesty Jr., a Conoco executive vice president, during the crisis.

Efforts to further regulate the industry intensified as the crisis continued. The Federal Energy Administration was formed in 1974, and it required oil companies to devote countless hours to complying with Byzantine reporting requirements. "The bookkeeping bureaucracy was enormous," says Gary Edwards. "We had to send the government all our cost records, including how we were applying cost pass-throughs to different classes of customers and different classes of trade and how our allocations were being handled [the government regulated how much oil each state and each customer could receive]. We had almost a company within the company just to comply with the regulations."

As the crisis reached the two-year point in late 1975, the U.S. Congress passed the Energy Policy and Conservation Act, which extended price controls for another 40 months. Congress also suggested that the integrated oil companies be split up into separate businesses, in effect requiring them to participate in only one phase — production, transportation, refining or marketing. Another proposal called for prohibiting petroleum companies from participating in other energy areas, such as coal and uranium. For a while, the government even considered nationalizing the upstream segment of the industry.

The U.S. oil industry reacted forcefully, asserting that vertical integration provided economic balance — a company's risks in one area were reduced by investments in another. Fragmenting the industry, on the other hand, would result in higher capital costs and, thus, higher prices to consumers. "The proposed divestiture legislation ignores the fact that the industry, as presently structured, is both competitive and efficient," Blauvelt said at the time.

Many economists agreed. "Any interruption of the flow of oil from wellhead to the consumer costs money," said David Schwartzman, professor of economics at The New School for Social Research in New York. "Integrated oil companies can keep the oil flowing because they own or manage production, transportation, refining and marketing. They can continuously balance crude oil supplies with shipping availability, refining capacity and consumer demand, so that fuels move smoothly and without interruption."

As these underlying economic benefits became more widely understood, the momentum to break up the petroleum companies lost steam. "When I look back, I realize it was nothing more than a threat," says Ralph Bailey, who succeeded Blauvelt as Conoco's chairman and CEO in 1979. "The government was reacting to public pressure. But bluster or not, in those days we didn't second-guess anything."

For Conoco employees and their families, the anti-oil industry fervor took a personal toll. In *Conoco* magazine, they reported their frustration. "There are moments when I'm made to feel almost as if we are outlaws," said the wife of a Conoco geophysicist. An offshore drilling superintendent said, "No more golf for a while. The other guys in my foursome preside over me like an inquisition."

E. J. "El" Grivetti, executive vice president of North American upstream operations when the crisis hit, says he could not go to a party without hearing somebody insult "those damn oil companies." "They were convinced we were holding back the oil," he adds. "The media was partly to blame. I remember this one outrageous story where it was reported that the industry had all these oil tankers off the coast of New Jersey waiting for the prices to come up before coming to shore. When you have a story like that, even though it's not true, the public is just dying to believe it."

Like other oil companies, Conoco fought the misinformation with facts and figures. "We got in front of the public with an aggressive public relations program," Bailey says. "Every executive, indeed every employee, was encouraged to learn the facts and to speak out forcefully against the tide of lies and misrepresentations."

The shortage made increased domestic and international exploration an urgent priority for U.S.-based oil companies. The run-up in prices gave Conoco and other oil companies the cash they needed to further this objective. Blauvelt used some of these revenues to reduce the company's debt-to-capitalization ratio from 40 percent to 32 percent, but most of the money — $5 billion — was poured into exploration and production. The primary location for this largesse was the inhospitable waters — but friendly political climate — of the oil-rich North Sea.

In the mid-1970s, Conoco was active in the North Sea, producing gas from several southern fields and oil and gas in the north. Its substantial interest in the Statfjord field was complemented by smaller interests in the Thistle and Dunlin fields — all in northern waters. Yet, despite Conoco's definite presence in the region, it had not yet distinguished itself as an operator of large offshore fields.

In 1975, Conoco at last undertook the first North Sea oil field project in which it would be the operator, with full responsibility for constructing and installing the facility — the Murchison project, named for a roguish nineteenth-century Scottish geologist, Sir Roderick Murchison.

The total cost to develop the field — $1.2 billion — dwarfed any other project the company had directed. By comparison, the initial cost to bring the Humber refinery on-stream in 1969 was a paltry $120 million. "How many times do you get a job where you can spend a million dollars a day for three years?" says Harry Sager, Conoco's manager of operations for Europe in 1976. This field, too, spanned an area owned by both the United

Conoco's development of the Murchison field marked a turning point in North Sea petroleum history. The field lay beneath 500 feet (152 meters) of storm-tossed seas, near the water depth record for the period. It involved U.K. and Norwegian partners and governments, requiring delicacy in the complex negotiations creating the project. It was Conoco's first oil development in Europe as an operator. Conoco's team installed the towering 866-foot-tall (264-meter) platform, left, and the steel substructure pictured on the following page amid the worst storms in decades. At $1.2 billion, the project was under budget. The platform began delivering oil to the Shetland Islands on schedule in 1980. Murchison incorporated many technical advances, including remotely accessed subsea wells, and propelled the company to an industry-leading position in project management.

CRISIS AND COMPETENCE

Kingdom and Norway, requiring delicate, high-level discussions with each country's political leaders. Conoco's partners in the British sector were British National Oil Corporation and Gulf — each with a one-third interest. The Norwegian portion was divided among nine partners, including Conoco Norway Inc. (Conoco's Norwegian operating company) with 10 percent. Together, Conoco's U.K. and Norwegian production interests in the Murchison field worked out to about 28 percent.

An internal organization of about 70 Conoco employees designed and directed the project, supplemented by some 550 employees of outside contractors. Over the course of the next four to five years, until the facility came on-stream in September 1980, they toiled to bring the project in on schedule and within budget, succeeding on both counts — an industry first. "That was a matter of great pride and accomplishment to all of us," Bailey recalls. "It was the most successful development in the North Sea at the time in terms of project development and technological innovation."

The Murchison field was in about 500 feet (152 meters) of water, coming within a few feet of the North Sea's record at the time for water depth, held by the neighboring Thistle field. The project comprised a single steel platform with 16 modules, three subsea wells and a 10-mile (16-kilometer) pipeline to the Dunlin field, where connection was made with the Brent pipeline system to a terminal in the Shetland Islands. The platform towered more than 44 stories above the sea, but more than half its 866-foot (264-meter) total height was underwater. "It established a unique reputation for Conoco as a successful major offshore project manager," says retired executive Dennis Gregg. "The project team was outstanding, led by Tom Marr (who earlier had been the project manager for the Viking gas field project), Jose Rodriguez, Dave Bowler, Gordon Hitchings, P. K. Lauren and Hank Hart. This team invented the concept of project services contractor — Bechtel, for Murchison — with a parallel organization providing big project know-how and systems."

"True to its reputation for daring," wrote *Forbes* magazine in 1979, Conoco "jumped into the North Sea with both feet."

By 1980, Conoco was producing crude oil from four North Sea fields — Murchison, Dunlin, Thistle and Statfjord. The Humber refinery processed the crude for sale as gasoline and distillates throughout the company's European retail network, which now included ARA, a Swedish marketing group the company acquired in 1971, in addition to its stations in Austria, Belgium, Italy, Luxembourg, the United Kingdom and West Germany.

The North Sea was not Conoco's only area of major exploration and production activity. The company's Advance Exploration Group, under the leadership of Dr. Richard Wing, guided teams of explorers to many points around the world. Indeed,

JEAN-MARIE PRIEUR

It's the great irony of his life that Jean-Marie Prieur was born in the middle of France and, though he dreamed of life as a sailor, didn't see the sea until he was 16 years old. Prieur, chief engineer of well operations for Conoco (U.K.) Limited, led a team working in the West of the Shetland Islands. He has been stationed in Aberdeen, Scotland, overlooking the North Sea, for 10 years. "When I look out across the water, I think about all the Conoco projects dotting the sea between me and my neighbors in Norway," he says. Prieur has worked on several, including the Hutton tension-leg platform, "an incredible technological breakthrough." He joined the company in 1991, following a stint with Shell Oil. "I knew this was the company for me. I feel I can get things done here. I have met and made the best friends of my life in this company."

Conoco held rights to more than 60 million acres in 19 countries by 1977 and was a partner in 33,000 producing wells worldwide. From these, Conoco obtained approximately 510,000 barrels of oil a day and 1.2 billion cubic feet of natural gas. In all, the company operated upstream and/or downstream facilities in 34 nations on six continents.

Another particularly promising oil-producing region was Indonesia. Conoco had received an interest in an offshore block in Indonesian waters in the late 1960s. In early 1978, after a decade of involvement in Indonesian exploration, the company signed a 30-year production-sharing contract with the Indonesian state oil company to explore in the northwestern peninsula of Irian Jaya. The following January, Conoco began producing oil from the Udang field, some 700 miles (1,127 kilometers) north of Jakarta. Its production platform, the largest in the country, was located in 300 feet (91 meters) of water, a record depth for Indonesia. A pipeline was built and a 93,000-deadweight-ton tanker, the *Udang Natuna*, readied. Production peaked at 50,000 barrels a day. "Indonesia was the most successful project our group had undertaken up to that time," says retiree John W. Strickland, Conoco's chief geologist in the Advance Exploration Group.

"We studied all the available literature, tried to find out if wells had been drilled previously and just generally put together bits and pieces of the puzzle to rate the area in terms of whether or not we should make an investment," he says. "We gave it the go-ahead, and the company took our direction." Indonesia remains a major upstream area for Conoco today.

In the United States, Conoco invested $84.4 million in the first federal lease sale of tracts in the Atlantic, off the coast of New Jersey. At first, development of the area, called the Baltimore Canyon Basin, was halted by a lawsuit brought by environmental interest groups who protested any offshore oil development. When the Supreme Court refused to intervene in the case in 1978, the oil industry was given a green light to begin drilling. Unfortunately, the Baltimore Canyon was a major disappointment for all parties in terms of finding any significant oil deposits, and Conoco was forced to write off much of its investment.

The Gulf of Mexico, however, provided better possibilities. Conoco continued drilling off Louisiana. In 1975, its semisubmersible rig, the *Pacesetter II*, earned the distinction of drilling in the deepest waters in which a commercial discovery had been reported in the Gulf. "This was only the beginning as far as what the Gulf of Mexico would mean to this company," says Ted Davis, vice president of exploration and production for North America at the time. "In fact, the Gulf would be the technological proving ground for all our future overseas deepwater drilling efforts."

As for its onshore U.S. production strategy, Blauvelt unleashed an all-out company effort to utilize new technologies to stimulate production in Conoco's older fields. Discovery wells that were tapped during the mid-1920s in such areas as the Permian Basin in Texas were redrilled and reworked to encourage more flow from their now-sluggish reservoirs. Altogether, Conoco spent some $300 million from 1972 to 1977 to redevelop many of its old fields, often with promising results. "By dipping more straws into the reservoirs, we were able to halt the annual 19 percent decline in our U.S. oil production," said John Whitman, Conoco's Midland division manager in 1977.

Conoco's seven U.S. refineries — in Ponca City, Oklahoma; Lake Charles, Louisiana; Denver, Colorado; Billings, Montana; Wrenshall, Minnesota; Paramount City, California; and Santa Maria, California — processed its U.S.-produced oil along with imported crude. New capacity was added via the purchase of the Sequoia refinery in Ponca City, bought from Cities Service Co. early in the 1970s and later consolidated into the existing Ponca City refinery. Each of Conoco's refineries also was modified in the mid-1970s to comply with U.S. Environmental Protection Agency regulations mandating industry development of a lead-free gasoline with 91 octane or better.

The expansion and upgrade of the Lake Charles refinery and the absence of a deepwater port ushered in Conoco's "lightering" operations, in which

BETTY J. CHRISTMAN

Betty Christman loves her job as an administrative coordinator at the Conoco refinery in Billings, Montana — and not just because she met her husband of 25 years there, G. D. "Corky" Christman, a retired district sales manager. "Since he started with the company first, we joke that he's branded with the Conoco triangle and I have a tattoo of the Conoco capsule." Christman, easygoing and warm, processes medical invoices, personnel paperwork and salaried payroll records and provides general administrative support. "When employees retire is when I really contribute to the life of this company," says Christman, "helping them and their spouses understand their benefits. In the event of a serious illness, I work with them to minimize delay and misunderstandings in handling the paperwork. My family only buys Conoco. My dad says, 'Betty, you can save two cents if you go down the block.' It doesn't make any difference. We're Conoco, through and through."

148   125 YEARS OF ENERGY

smaller tankers received crude oil from VLCCs (very large crude carriers) at sea for redelivery at the Port of Lake Charles. The company purchased three ships for lightering — the 40,000-deadweight-ton *Texas* and the 60,000-deadweight-ton *Oklahoma* and *Louisiana*, which received oil from four 270,000-deadweight-ton VLCCs the company purchased in the 1970s — the *America*, the *Canada*, the *Europe* and the *Independence*. The lightering vessels also delivered oil at Freeport, Texas, for transportation to Conoco's Ponca City refinery via the Seaway pipeline.

Conoco's marketing philosophy during the decade reflected the shift in consumer preferences to more choice in the types of gasoline and methods of service provided. Recession-weary motorists made it clear they preferred to save a few cents a gallon on gasoline and do without such frills as attendants, windshield cleaning and oil-level checks. The company unveiled its first self-service gas stations in Houston and gradually expanded the strategy to other areas. It also withdrew from certain markets and converted many company-owned sites to low-priced marketing under the FasGas or E-Qual brands.

The company simultaneously began a concerted program of reducing its marketing properties. In the summer of 1973, before the oil embargo reshaped the American petroleum business, Conoco was selling gasoline through more than 7,500 stations, more than half of which were wholly owned. The profusion of regulations that followed the embargo forced the company to take a hard look at its wholly owned stations. It found that many were producing a low rate of return, sometimes less than 1 percent annually on the money invested.

"Our strategy is to hold on to those stations that are producing better-than-average returns, and offer those that aren't to jobbers and dealers at a fair price," said Bruce McCall, vice president of North American petroleum marketing in the mid-1970s. By 1978, the company had disposed of some 1,500 stations and bulk plants, or about 60 percent of its domestic marketing assets. The properties were sold for close to $100 million altogether.

Conoco's employees during the 1970s included many baby boomers just beginning their careers. A hardy lot, they had endured the tragedies associated with the war in Vietnam and were now beset by

Conoco's Lake Charles, Louisiana, refinery, opposite, was modified in the mid-1970s to produce lead-free gasoline. The expansion and upgrade of the Lake Charles refinery and the absence of a deepwater port ushered in Conoco's "lightering" operations, in which smaller tankers received crude oil from VLCCs (very large crude carriers) at sea for redelivery at Lake Charles. Pictured above is the lightering vessel *Louisiana*.

CRISIS AND COMPETENCE 149

More than 44,000 Conocoans gathered around the world in 1975 to celebrate the company's centennial. Together with their families, Conocoans in Ponca City, above, enjoyed a carnival and exclusive concert.

both a national recession and a global energy crisis. Nevertheless, more than 44,000 stalwart Conocoans of all ages gathered in distant parts of the globe in 1975 to celebrate the company's centennial. Many assembled in Houston at the Woodlands Inn for a daylong bash. Swimming, boating, riding horses or listening to the musings of a roving fortuneteller, they put behind them the trials of the ongoing recession and energy crisis. It was a rare opportunity for all employees to take pride in the company's remarkable accomplishments and ponder the many challenges ahead.

Ralph E. Bailey was elected president of Conoco in November 1977, filling a position that had been vacant since Blauvelt eliminated the presidency during the reorganization of the company two years earlier. In 1979, following Blauvelt's early retirement, Bailey assumed the chairman and CEO responsibilities as well. Ruggedly handsome and a charismatic leader, Bailey sealed the personality rifts that had erupted during Blauvelt's rather stern reign.

Bailey, a mechanical engineer who rose through the ranks of the coal industry to become president and CEO of Consol, was more of a humanist. "He was a great leader," says Archie Dunham. "He had an ability to relate to all employees of the company in a very open, gracious, nondefensive manner. Yet, he was clearly the CEO, tough as nails and very quick. His confidence helped him be a very natural leader." Gary Edwards recalls that Bailey had the uncanny ability to remember names of employees, even though he had met them only once or twice. "It was a very personal touch," Edwards says.

Blauvelt left the company in sound financial shape, but its earnings had barely kept pace with inflation. The $5 billion he poured into exploration and production failed to produce the hoped-for results on the income statement because of the hostile political environment. "Howard Blauvelt's problems were external more than they were internal," *Forbes* reported in 1979.

The political situation had not improved much when Bailey took the wheel. A long labor strike devastated Consol in the first quarter of 1978, and the company lost $52 million. At the same time, the oil industry was being buffeted by Act II of the Middle East oil crisis, brought about by the overthrow of the Shah of Iran in 1978 and 1979. Conoco had two assets in Iran: a small ownership in the Iranian International Consortium (IRICON), acquired when it purchased San Jacinto, and a Conoco-operated license in the Bandar Abbas area, on which major natural gas reserves had been discovered but not developed.

Conoco's position in Iran was formative. The IRICON crude oil volume was relatively small and simply sold on world markets. Conoco's large discovery at Bandar Abbas had potential but no immediate importance because of a lack of markets. Nonetheless, these assets were lost to nationalization when the Shah went into exile, and the entire industry suffered the impact of the price increases in the aftermath of the Iranian revolution.

Many of the onerous regulations enacted in the wake of the 1973 embargo were still in effect, including price controls, allocation limits and exploration constraints. New regulations were adopted which extended natural gas price controls to intrastate gas. Public animosity returned, and the oil industry again was vilified. Indeed, the number one television show in the nation was *Dallas*, an hour-long series that dramatized the "evil manipulators" supposedly running the nation's oil companies.

As foreign crude supplies dwindled, the long lines reappeared at U.S. gas stations. Conoco again sent out its executives to explain the facts behind the shortage. The company's Conoco Speakers program encouraged all employees to give as many speeches as possible, and many Conoco retirees and current employees recall dozens of speaking engagements, television appearances and press interviews. The speeches numbered in the thousands.

In the summer of 1979, the motoring public braced itself for the shortages. Gas lines in some areas became organized institutions, with some motorists acting as scouts to report on the queues' progress and others taking orders for coffee and breakfast rolls. Enterprising teenagers of driving age turned the event into an opportunity to make a few bucks — sitting in neighbors' cars in the lines while they got the kids off to school. In short, the public made the best of it.

By August, the long lines vanished as quickly as they had appeared. The Arab oil-producing countries increased imports to the U.S., and the oil companies turned them into products as quickly as possible. The unsung heroes, however, were American motorists. Many parked their cars at

In the 1970s, Conoco established the Conoco Speakers program to provide its perspective on critical energy issues to a distrustful public. The Golden Gavel was awarded to those employees making 10 presentations or more a year. In 1977 alone, more than 400 employees made over 2,900 speeches and presentations to civic, business and educational audiences, such as the high school students above.

CRISIS AND COMPETENCE 151

the curb during the shortage, a conservation effort that added up to a 9 percent reduction in demand over the summer of 1978. "Without it, we would have been in trouble," said Bruce McCall.

The crisis encouraged a renewed emphasis on the development of synthetic fuels. In September 1979, U.S. President Jimmy Carter announced an ambitious but unrealistic plan to make 2.5 million barrels of synthetic fuels available daily by 1990. Conoco rose to the occasion, devoting most of its alternative fuels research to coal gasification, shale oil and ultra-heavy oil. In 1981, Conoco and Chevron tested the feasibility of methanol (produced from coal) as a gasoline substitute by driving a car fueled only by methanol all the way across the country. Although these efforts were scientifically successful, they never proved commercially viable. "The technology was there, but it just never panned out economically, and within a few years, the whole thing dissipated," Bailey says.

Another initiative, encouraged by the downstream technology group, was the development of fluidized bed combustion technology, an effort to more efficiently utilize the company's fuel coke. "The technology made possible burning higher-sulfur coal and petroleum coke in an environmentally friendly way by injecting limestone to react with the sulfur," says Don Butter. The process was used a few years later at Lake Charles, where an existing power plant was retrofitted to allow the burning of petroleum coke being produced in Conoco's refinery. "This same technology," says Butter, "may eventually provide a market for petroleum coke out of Venezuela."

While the second Middle East crisis reverberated, Conoco expanded its hunt for domestic oil reserves. The company signed a partnership agreement in 1978 with E. I. du Pont de Nemours & Company for a five-year joint oil and gas exploration program. The agreement called for any future reserves discovered by the partnership to be owned two-thirds by Conoco and one-third by DuPont. Like other industrial energy-users, DuPont was concerned about the impact of fluctuating oil and gas availability on its operations and was determined to become more energy self-sufficient. In effect, it wanted more control over its destiny.

Little did either company know it at the time, but their partnership, and the relationships it engendered, would play a major role in Conoco's destiny.

In the spring of 1981, an enterprising Canadian oil company, Dome Petroleum Limited, made overtures to Conoco about acquiring Hudson's Bay Oil and Gas Company, the Canadian company E. W. Marland had created decades before. It was the beginning of what would become Conoco's greatest challenge and a pivotal chapter in the history of American business.

"Dome approached us with a tender offer for

Pictured at left and opposite is Conoco's research and development facility in Ponca City in the 1970s, where Conoco worked on alternative fuels research. Computers, above, were playing a bigger role, not just in the research process, but also allowing offshore production sites to be controlled from shore bases. Nearly all of Conoco's production in the Gulf of Mexico was computer-controlled, far more than that of any other operator.

Conoco shares," Bailey recalls. "Their plan was to trade the shares back to us for our 51 percent ownership in H-BOG. The tender was unfriendly and made with only a $10-a-share premium over the market value. We turned them down." In doing so, Bailey unwittingly opened Pandora's box, setting in motion a series of speculative maneuvers that ultimately led to its acquisition by DuPont — the very first of the many mega-mergers that captivated the business world in the 1980s.

After being rebuffed by management, Dome made an offer to Conoco's shareholders, tendering for 20 percent of Conoco's shares, the amount it figured would be needed to trade for H-BOG. (No other single block of shares in the company at the time accounted for more than 2 percent of Conoco.) In May 1981, Dome acquired the 20 percent block of shares it was seeking. But Conoco shareholders' response to Dome's offer was something that nobody had expected. Approximately 60 percent of Conoco's outstanding shares were tendered, leading observers to believe that full control of Conoco could be bought at the per-share price Dome was offering.

Bailey realized the company was now in play and vulnerable to a full takeover. "The wolves began to pounce," he says. The first overture came from the U.S. subsidiary of The Seagram Company Limited, the Montreal-based company. Bailey reacted by negotiating a deal with Dome giving it H-BOG in return for Conoco stock and cash.

The maneuvers and machinations of Conoco's many suitors — a group that included Seagram, Mobil, Texaco, Marathon Oil and Unocal — captivated the business press and the major national papers. On September 30, 1981, E. I. du Pont de Nemours & Company acquired Conoco at a cost of $7.4 billion.

With Dome's cash in hand, Bailey approached Cities Service Co. with the offer of a merger. On the very day the boards of the companies agreed, the information leaked out. Seagram came knocking with an unfriendly cash offer worth about $2.5 billion, for what would amount to a 40 percent interest in Conoco. The merger agreement with Cities Service was called off.

Within days, however, more suitors lined up, including Mobil, which proffered an unfriendly offer, and Texaco, with a friendly one. Conoco's board rejected both. Marathon Oil and Unocal also expressed interest in the company. A bidding war ensued, compelling Seagram to increase the value of its bid. "We were faced with a situation where Seagram had assembled a very large block of stock through their tender offer to shareholders and had cleared all the regulatory hurdles," Bailey says. "We looked like a goner."

The company mounted the best defense it could. Aware that regulations forbade a liquor manufacturer from owning a retail liquor company, Conoco contemplated acquiring a liquor retailer. "I was working for Dino Nicandros at the time, and we were looking at any strategy that would keep us independent," Dunham says. "I remember calling our regional marketing reps and saying, 'You don't know me, but we're working on a strategy to save our company, and I need you to go to Jake's liquor store on the corner of Main Street and 8th and prepare an estimate of what you think that store is worth.' Of course, we had no experience in this regard, but we were frantic."

Bailey, meanwhile, set about to find a white knight to save the company from Seagram's takeover. "It had to be a company that would be a more satisfying match for Conoco than Seagram and a better deal for our shareholders to boot," he says. "Such a company would need to be financially strong, highly regarded on a global basis and, most importantly, not another oil company that would simply merge us out of existence."

That company was its U.S. oil and gas exploration partner, DuPont. "I personally approached Ed Jefferson [chairman and CEO of DuPont at the time], and he indicated they were ready to go," Bailey says. "The two of us sat down with our investment bankers and, with both our boards' approval, cut a deal." The terms, approved on July 14, called for DuPont to acquire 100 percent of Conoco's stock in a transaction worth about $7.3 billion, making it the largest merger deal in U.S. history. Most importantly, the transaction provided for Conoco to operate under its current management as a wholly owned DuPont subsidiary.

Before the ink was dry on the deal, other suitors increased their bids. In response, DuPont upped its ante to $7.4 billion. The business press carried the maneuvers and machinations virtually every day, often on the front page. Hundreds of op-ed columns and political cartoons expressing a plethora of views were printed. One cartoon, in particular, shows a rather frustrated tattoo artist hovering over a patron (Conoco) adorned with several tattooed hearts that have been crossed out. Inside each heart, Conoco is linked with one of its many suitors.

The corporate bidding war finally ended — six weeks after the merger announcement. The U.S. Department of Justice gave its blessing, and on September 30, 1981, Conoco became a wholly owned DuPont subsidiary. A visibly exhausted Bailey rejoiced. "The entire process was excruciating," he says today. "Every morning I'd go to the office and look at what I called the 'chess board.' Each day, the pieces moved this way and that, indicating a new direction taken by a particular bidder. Meanwhile, their every move — our every move — was being watched by the world. Fortunately, I didn't worry alone. I had a great team of operating people that fought the good fight with me."

Weighing on his conscience the entire time, Bailey says, was the dire need for Conoco to remain intact. DuPont gave him that assurance. "I always felt that if I could somehow keep Conoco together, it would be sprung loose someday. Ever since then, I have kept my hopes high that the flag emblazoned with Conoco's red capsule would again fly as an independent."

Seventeen years later, Ralph E. Bailey's fondest wish would come true.

Realignment and Refinement

Bloodied but unbowed after months of fending off unwanted suitors, Conoco pondered its new future as a DuPont subsidiary. Once before it had been part of a much larger organization — John D. Rockefeller's Standard Oil Trust. This time, its new partner would give it far more autonomy. "What did we win?" Ralph Bailey rhetorically asked employees who gathered on August 5, 1981, to hear the chairman's remarks on the acquisition.

"We won the right to run Conoco. We won the right for this organization to be kept together. And we won the right to carry out our plans and dreams." Bailey received assurances from DuPont's chairman, Edward G. Jefferson, that "they intend for this management to run Conoco just as it was run before," he said. "Not one single job would be in jeopardy. They recognized in the beginning that this was a successful management, that this company has a terrific future."

Bailey is credited with keeping the relationship between DuPont and Conoco cordial, stable and separate. "After most acquisitions, it's customary for the CEO of the purchased company to move on," says Rob McKee, executive vice president for exploration and production. "Ralph made a principled pact with himself that he would stay on to ensure that Conoco maintained its identity and got a fair shake out of the merger. He would not agree to any organizational moves by DuPont that would diminish power on the Conoco side. Strictly through force of personality, he got what he wanted."

The marriage had some early ups and downs. In July 1983, Conoco's corporate headquarters were moved from Stamford, Connecticut, to DuPont's Wilmington, Delaware, headquarters. The decision to house both companies together caused anxiety among Conoco executives. "We at Conoco have a philosophy of doing things sooner rather than later — quite different from DuPont's more methodical mentality," Dino Nicandros explains.

"Although we were financially focused, had sharp control of our investment returns and exemplified the risk-taking spirit that comes from being an oil company, DuPont, I'm afraid, saw us as a bunch of cowboys. They, on the other hand, were very gentlemanly, nonconfrontational, bureaucratic and hierarchical — truly old school."

Bailey appointed Nicandros chairman of Conoco's merger team. "I had a counterpart at

DuPont, one of the world's leading manufacturers of chemicals, is also one of the most innovative. Products like Kevlar, seen "brewing" in large vats to the left, and nylon, below, have transformed daily life. Opposite, Conoco CEO Ralph Bailey assured employees that as part of DuPont, Conoco was safe from the threat of takeover and free to operate under its own management. Conoco's petroleum reserves promised DuPont energy security in an uncertain energy environment. Energy and feedstock access were major factors behind DuPont's acquisition of Conoco.

DuPont [Executive Vice President David Barnes], and we worked together to implement a smooth transition," Nicandros recalls. Over time, the pieces fit together.

"Bailey had established early on that Conoco was a well-run oil company and should be left alone," Nicandros continues. "The merger committee institutionalized this by developing intracompany communications protocols. Anytime anybody from DuPont called someone at Conoco, they first had to talk to me. By doing this, we blocked multiple attempts by DuPont troops to tell us what to do and how to do it." Protocols or no, as the years progressed, Nicandros asserts, both companies learned from each other and became better for it.

The cost of the acquisition for DuPont compelled Conoco to refocus its business to reduce costs and DuPont's acquisition-related debts. Conoco pared its asset holdings in the United States, selling 29 oil- and natural gas-producing fields to the Petro-Lewis Corporation for $772 million in 1982. The following year, DuPont sold the lion's share of Conoco Chemicals for about $600 million in cash to a group of investors who renamed it Vista Chemicals, thus bringing to an end Conoco's four-decade involvement in the specialty chemicals business. In 1984, DuPont also sold Continental Carbon, or Concarb, to Witco Chemicals.

Conoco also rationalized its downstream organization. It sold Coastal Oil and Home Fuel Oil in 1981 and a year later consolidated its Western Stores marketing division into its Kayo Oil subsidiary based in Chattanooga, Tennessee. It closed the Wrenshall refinery in Minnesota and reduced crude oil inventories, saving significant working capital. In late 1982, the company sold the Paramount oil refinery in California (part of its Douglas Oil subsidiary) to Oasis Petroleum (not to be confused with Conoco's Oasis Oil Company of Libya). Eventually, the company's marketing operations were blended into the Conoco system, and the rest of Douglas Oil's assets were sold or discontinued.

Another major subsidiary also seemed headed for the auction block — Consol. "Ed Jefferson [chairman of DuPont] didn't care much for Conoco's coal operation," recalls Ike Krasts, an analyst in Conoco's economics division at the time and later vice president of planning. "He felt it had too much going against it."

Indeed, Consol was beset with problems. Its production included a large percentage of high-sulfur coal at a time when the government wanted coal-burning utilities to cut down on sulfur emissions because of "acid rain" concerns. Consol under Conoco had reduced its injury rate each year and led the coal industry in safety just as Conoco led the petroleum industry and DuPont led the chemical industry. But the industry's history — including a spotty safety record and a restive labor movement going back to John L. Lewis and the United Mine

---

Consolidation Coal Company (Consol) endured a difficult period during the first part of the 1980s. The company fought weak coal prices by switching to the production of low-sulfur coal, making its mines more productive and entering into long-term contracts with buyers. Consol also completed several expansions, including the development of the Bailey mine in western Pennsylvania. In this photo, a Consol worker prepares coal for shipment to an electric utility customer. In 1987, Consol became a wholly owned subsidiary of DuPont, ending Conoco's relationship with the company. DuPont later sold off its Consol shares.

THE HOUSTON COMPLEX

Conoco's Houston headquarters, completed in 1984, was designed by architect Kevin Roche. Roche based his design on the aristocratic structures built during the ancient Heian period in Japan, creating the effect of a natural oasis in a part of Texas subject to both extreme heat and torrential rain. The design incorporates aesthetics and amenities that provide a serene work atmosphere and a definite connection to the surrounding landscape. Sixteen three-story buildings are arranged around a central walkway connecting them on the second story. The campus-like setting includes a man-made lake, stocked with fish, that borders the buildings, some of which even appear as if they are floating on the lake. Ninety percent of the offices are of equal size, and most offices have a window providing views of the lake, greenery and jogging path. A cafeteria, travel agency, bank and medical facility are among the many other conveniences provided for employees. The headquarters complex is close to where many Conocoans live and is located in Houston's "energy corridor." The company needed room to expand, something the new facility — totaling more than one million square feet (100,000 square meters) of office

space — easily provides. Canopied parking areas are also provided free — a big-city perk offering economy and shade. Given the vastness of this architectural marvel, visitors are provided maps to navigate routes among the 1,800 offices. Many of the buildings encompassing the headquarters are named for a project or personage important to Conoco's history, such as the McCollum Building and the Dubai Building.

Workers in the 1920s — was not easy to overlook. Moreover, coal prices in the early 1980s were heading south. These various factors, together with DuPont's need to sell assets to raise cash, put Consol's future in jeopardy.

But such plans faced a formidable obstacle. "Ralph Bailey was a coal man, and there was no way he was going to sell the company he once headed," Krasts says. Bailey decided the best way to block the sale was to make Consol too good a subsidiary to let go. "He and Bobby Brown [Consol's chairman] got together and made a deal with U.S. Steel to buy their low-sulfur mines and, consequently, their long-term contracts with buyers," Krasts adds.

"They made the mines more efficient and productive, enabling Consol to supply coal at a much lower cost than U.S. Steel had. Thus, over time, Consol began to devote more of its business to the production of low-sulfur coal. Based on the cash flow Consol was expected to produce over the next 10 years, Jefferson was dissuaded from selling the company." Upon Bailey's retirement in 1987, Consol became a DuPont subsidiary. In 1998, having sold half of Consol eight years earlier, DuPont sold the remainder.

The slimming of Conoco's operations was complemented by management changes. In January 1983, Bailey announced that 49-year-old Dino Nicandros would succeed Mike Morris as president of Conoco's petroleum operations. Nicandros would oversee Conoco's oil enterprises from the company's new Houston complex.

The aim of the company in creating this new operations headquarters facility, completed in 1984, was twofold. Conoco Tower at Greenway Plaza was completely full, and the company needed room to expand. Moreover, the agonizing commute to Greenway Plaza begged a more conducive atmosphere for work and innovation — a building that would boost employee morale. The new facility did just that, providing airy and open office interiors set in an idyllic landscape filled with natural beauty. Overall, the unique structure produces the effect of a green oasis. Its campus-style plan encompasses more than one million square feet (100,000 square meters) of total office space, abundant walkways and multiple three-story structures that appear to float on a natural lake.

These positive developments were tempered by the difficult market conditions of the 1980s, caused by declining crude oil and gas prices. Conoco was forced to cut expenses wherever it could, including reducing the size of its work force. In 1985, the company offered an early retirement incentive package to older employees. For many, it was an opportunity to receive a significant lump-sum payout — based on the number of years of service — to pursue new adventures. For others, it was an expression that times had changed and the Conoco of old was no longer.

REALIGNMENT AND REFINEMENT 163

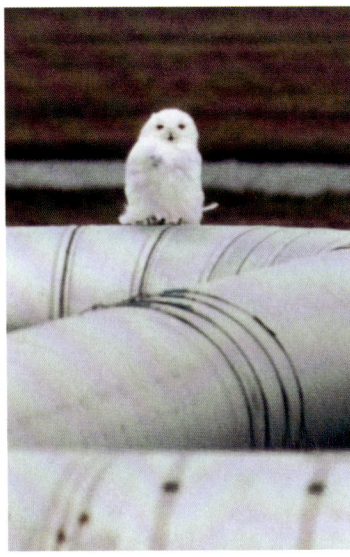

Conoco partnered with four other oil companies to begin development of the Milne Point field on the Alaskan North Slope in 1984. Conoco battled frigid Arctic temperatures while installing the facilities in the field and constructing a 14-inch pipeline from Milne Point to the Trans-Alaska pipeline. The field began producing in late 1985, but only months later, oil prices plunged, stalling further development. In 1993, Conoco traded its Milne Point assets to British Petroleum for a 33 percent interest in British Petroleum's Mississippi Canyon holdings in the Gulf of Mexico.

In the early to mid-1980s, a period of active exploration and technological innovation, Conoco searched for oil and gas in ever more inhospitable regions of the globe, from the barren tundra of the Arctic Circle to the deep waters of the Gulf of Mexico and the North Sea.

The government controls and red tape that had impeded U.S. oil exploration and development during the energy crises loosened considerably. On his first day in office, U.S. President Ronald Reagan, deeply concerned over American reliance on foreign oil, deregulated U.S. oil prices. The new president also encouraged a buildup of domestic reserves. In July 1982, despite a barrage of criticism from environmental groups, the U.S. Department of the Interior dramatically expanded the federal acreage available for leasing on the Outer Continental Shelf, much of it along the Alaskan coastline. Conoco headed north.

Oil had been discovered at Milne Point on the Alaskan North Slope in early 1980, and a delineation well was completed later that year. In 1984, Conoco partnered with four other oil companies to begin development of the Milne Point field and was named the operator of the project — the third operator on the North Slope. With oil selling at $35 a barrel — a sizable sum given that oil sold for a dollar a barrel in 1940 — optimism ran high that enough Alaskan crude could be produced to justify the high cost of production.

The challenges loomed large. The temperature was a bone-chilling 50 degrees below zero (minus 46 degrees centigrade) in the winter, which felt like 100 below (minus 73 centigrade) in the biting wind. The cold was a bitter enemy for the company's Alaskan personnel, many from Conoco's Continental Pipe Line Company, which was charged with building a 14-inch pipeline to carry Milne Point oil to the Trans-Alaska pipeline system. A blinding "white-out" could occur if the wind whipped up the snow, and it was easy to become disoriented. Everyone carried two-way radios to keep in touch with the central facilities office, where people and equipment were tracked constantly.

"The production technology we used at Milne Point actually was not very different from what we were using at other oil fields," recalls Rob McKee, then Conoco's western regional manager of production. "But, because of the distance from civilization, very sensitive ecology, hostile weather and the size of the field itself, the costs at Milne Point were much greater." In fact, on the North Slope, the capital investment for accommodations alone averaged $500,000 per person.

In late 1985, oil flowed for the first time from the Milne Point field to the Trans-Alaska pipeline. The project had cost Conoco and its partners more than $325 million, with Conoco chipping in close to $175 million. At the dedication ceremonies, Alaska's governor toasted the success of the project, noting how difficult the journey had been.

REALIGNMENT AND REFINEMENT

Conoco built the world's first tension-leg platform, or TLP, for the Hutton field, 90 miles (145 kilometers) northeast of the Shetland Islands. Unlike conventional fixed production platforms, which are positioned above sea level by legs that rest on the seafloor, the TLP floats on the surface — moored vertically to the seabed by tubular steel legs, pictured above.

Everyone was in high spirits.

The enthusiasm was short-lived. In early 1986, oil prices crashed. Conoco kept Milne Point running for a year — hoping oil prices might recover — but eventually made the difficult decision to suspend production early in 1987. Although operations were resumed for a short time in 1989, oil prices never recovered sufficiently to salvage the project. "We built Milne Point predicated on oil prices being in a minimum range of $25-28 per barrel," Harry Sager says. "With oil selling at $15 a barrel, we couldn't yield a desirable return on our investment."

In 1993, Conoco traded its Milne Point assets to British Petroleum for a 33 percent interest in British Petroleum's Mississippi Canyon holdings in the Gulf of Mexico.

Many successful projects were undertaken in the early 1980s, several of them focused on the North Sea. After a long industry hiatus, Conoco once again began exploring and developing the Southern Basin area of the North Sea. The company was eager to service the interests of British Gas, which in 1981 indicated a desire to bolster reserves.

The first project Conoco brought on-stream was the Victor field development. The project team, led by John Barnes, adhered to an aggressive schedule to secure a large bonus from British Gas. The Kotter and Logger fields in the Dutch sector of the North Sea were also exceptionally well-managed projects in which Conoco participated and were the site of industry records, such as the heaviest offshore lift. Conoco participated in many other successful projects in the North Sea, including Miller in the U.K. and Statfjord C in Norway.

The North Sea, in fact, was the site of one of the industry's most important technological breakthroughs. In 1980, Conoco announced it would build the world's first tension-leg platform, or TLP, for the Hutton field, 90 miles (145 kilometers) northeast of the Shetland Islands. The field, in itself, was nothing remarkable, with recoverable reserves estimated at about 200 million barrels of high-quality crude, qualifying it for medium-size status in the grand hierarchy of North Sea fields. What made the Hutton field revolutionary was the TLP, which would permit production in vastly deeper waters.

At the time, drilling and production in 600 feet (183 meters) of water approached the outer limits of technological possibility. "We had to find a way to bridge the gap between drilling in deep waters and bringing up the oil for production," says Buck Curtis, the father of the TLP and the Khazzan underwater storage tanks of a decade earlier.

The TLP was the solution. The concept derived from government studies during World War II concerning the feasibility of building floating airfields that could be tethered to the mid-Atlantic floor. "Back in the 1970s, Hal Nabors, the president of our Dubai Petroleum Company, asked me if I thought a floating production platform leashed to the seabed had any

The TLP concept derived from government studies during World War II concerning the feasibility of building floating airfields that could be tethered to the mid-Atlantic floor. As seen in the model of the Hutton platform at left, the buoyancy of the platform keeps a constant tension on the tethers anchored to the seafloor, preventing any upward or downward movement of the platform and allowing a maximum horizontal shift in any one direction of just 25 yards (23 meters). Commercial application of the TLP concept was the brainchild of Buck Curtis and earned him DuPont's prestigious Lavoisier Medal. It was also Conoco's first successful endeavor in its ongoing search for new methods to economically develop oil and gas fields in very deep water. On the following page, the Hutton platform is seen being lowered into the water.

merit," Curtis recalls. "I said, 'Hmmm.' So we got our research people together in Houston and I led them in the development of the project."

Curtis gives much of the credit for the TLP's design to Conoco's outstanding exploration and production technology department. Unlike conventional fixed production platforms, which are positioned above sea level by legs that rest on the seafloor, the TLP floats on the surface — moored vertically to the seabed by tubular steel tethers, or legs, that hold the platform below its normal buoyancy level. This creates an upward force that keeps the legs under constant tension. Although the platform can move horizontally with waves, winds and currents — a maximum of 25 yards (23 meters) from its center position, in extreme weather — it cannot move up or down. And, since a TLP is built and assembled onshore in sheltered waters, there's a lot less hardware, time and effort expended to assemble it on location in the field. "When weighed against the prospect of building bigger and bigger underwater skyscrapers, the TLP was extremely affordable and attractive," Harry Sager says.

The Hutton field lay in only 480 feet (146 meters) of water, but it was deep enough for the TLP concept to be tested. On July 15, 1984, the TLP was christened. "Mrs. Thatcher, the prime minister, invited senior Conoco management to a reception at 10 Downing Street to mark the event," recalls Dan McGeachie, director and

The Hutton TLP, illuminated above, was a stunning technical achievement that provided new sources of energy to the world. This brilliant concept was honored by the U.K. government with a commemorative postage stamp. British Energy Minister Peter Walker christened the historic facility, and British Prime Minister Margaret Thatcher marked the event with a special reception.

general manager of government and public affairs for Conoco U.K. "And the BBC made an hour-long, award-winning TV program about the project." Oil flowed at a rate between 75,000 and 90,000 barrels a day.

Although the project ultimately cost Conoco more than $1.3 billion, it put the company at the forefront of deepwater technology. "The TLP was a PR person's dream. The world's media were practically camped on our doorstep," recalls Sondra Fowler, manager of public affairs. "The *New York Times* gave it a whole page in the science section, *Popular Mechanics* did a center spread, and through the influence of Dan McGeachie, the U.K. government created a TLP postage stamp."

Despite its relatively modest size, Conoco was continuing to shake up the industry with groundbreaking technology. At the 1985 Offshore Technology Conference — the foremost international forum for technology affecting offshore operations and development — Conoco (U.K.) Limited was awarded the Distinguished Achievement Award for organizations, recognizing the success of the tension-leg platform project. At the same conference in 1987, Buck Curtis won the award for individual lifetime achievement.

The TLP proved that oil could be produced economically in water depths beyond the limits of conventional fixed structures. For the first time, drilling and producing oil in 2,000 feet or even 3,000 feet

Conoco set the world's record for deepwater oil production at 1,760 feet (536 meters) of water with the development of the Jolliet field in the Gulf of Mexico, above, in 1989 — a distinction reflected in this Conocoan's beaming smile.

(more than 900 meters) of water — unimaginable a decade earlier — was within the realm of possibility. "Hutton opened the door for everything else," Curtis says.

Whatever the scientific wisdom of the day that said "no" to deeper drilling, new kinds of oil platforms or just new ways of doing things, Conoco's scientists and engineers said "maybe." The company took this success a step further by finding new applications for its innovations. The TLP concept fostered similar structures in other parts of the world, including a tension-leg well platform, or TLWP, in the Gulf of Mexico's Green Canyon area — about 150 miles (241 kilometers) south of Morgan City, Louisiana. In December 1982, Conoco engineers probing the basin with the *Lexington*, a semisubmersible drilling rig, had drilled a well in 1,835 feet (559 meters) of water — a record depth. The water was so deep that "the Eiffel Tower could be stacked twice, leaving barely 100 feet to poke above the surface of the water," *Conoco World* reported at the time. Although the drilling tests in what became known as the Jolliet field had proved the presence of hydrocarbons, the feasibility of producing oil at such depths was in doubt.

The success of the Hutton project erased those doubts. On July 15, 1989, five years to the day after the installation of the Hutton TLP, Conoco completed the Jolliet TLWP in 1,760 feet (536 meters) of water. The TLWP was a scaled-down version of the TLP — one-fourth the size of Hutton — with minimal production, processing and accommodation facilities. The TLWP's production capacity also was leaner, averaging about 35,000 barrels of oil and 50 million cubic feet of natural gas per day.

When the first oil flowed from Jolliet in late 1989, Conoco had the distinction of setting the world water-depth record for oil production from a platform. Remarkably, only a decade later, Conoco would share in another Gulf record with the installation of the Ursa TLP, a production platform in 3,800 feet (1,158 meters) of water — more than twice the depth of Jolliet.

By 1986, Conoco was one of the largest operators of platforms in the Gulf. Nearly all its oil and gas production was computer-controlled from shore bases, far more than that of any other operator. "We put automation control systems both onshore and offshore so that during hurricane evacuations we could bring people out of harm's way, yet keep the cash register running," says Brent Meyers, manager of upstream information management. "Since then, we've run these offshore facilities without shutdown during several hurricanes."

Not everything was smooth sailing, and, in fact, the company experienced its share of failures in the 1980s and early 1990s. Major exploration plays failed in Egypt's Gulf of Suez, the difficult mountainous jungle of Irian Jaya and the West African countries of Gabon, Congo and Angola. "Small oil

The foundation template for the Jolliet tension-leg well platform (TLWP) is lowered into place on the seafloor. Jolliet incorporated technology from earlier TLPs but advanced and simplified many of the key elements. These new designs positioned Conoco as a leader in deepwater production technology.

REALIGNMENT AND REFINEMENT 171

During the 1980s, gasoline price wars in Conoco's European markets, particularly Germany, Belgium and the U.K., compelled the company to restructure its portfolio. Conoco's Italian retail operations and many under-performing stations elsewhere in Europe were sold. The company fought to establish a niche as a low-price retail marketer under the Jet and Seca brands, making a significant investment in architectural improvements and modernization of its retail outlets, such as this sleek Seca station in Belgium.

discoveries were made in the Gulf of Suez and Congo, but they didn't make the grade commercially," says Dennis Gregg. Significant discoveries were made in the rain forest of Ecuador, but Conoco did not consider those fields sufficiently viable for development due to marginal commercial terms and strong opposition from a small number of environmental activists.

However, the project team engineered innovative ways to deal with the sensitive rain forest ecosystems and the primitive people living there. "In the final analysis, several of the creative programs to minimize the impact on these environments enhanced the company's position as an industry leader in environmental care and safety," Gregg adds.

Downstream operations during the period were updated in several areas. Importation of heavier (and higher-sulfur) Venezuelan, Mexican and Middle Eastern crudes demanded more sophisticated refinery methods and equipment. In late 1983, Conoco completed a $200 million modification of its 42-year-old Lake Charles refinery to process heavy, sour (high-sulfur) crude oil. New equipment included a delayed coker, two desulfurizers, a sulfur plant and a gas recovery plant. A rotary calciner to produce petroleum coke was installed in 1986. New marine terminals for crude oil and coke were added along the Calcasieu River. As the Lake Charles refinery became capable of processing high-sulfur crude oil, Conoco turned from lightering vessels to tankers, including the original *Constitution* and the *Sentinel*.

"The sour crude conversion is a major step in our efforts to lessen our dependence on sweet (low-sulfur) crudes, which are becoming scarce and more expensive," said Ozzie Newell, Conoco's executive vice president for refining at the time. "We must be able to draw on crude supplies from all possible sources."

The Humber and Ponca City refineries also underwent improvements in the mid-1980s. An $80 million fluid catalytic cracker was installed at Humber in 1986. The equipment augmented the ability of the refinery to meet increased requirements in Europe for low-lead gasoline.

In the 1970s and 1980s, Conoco's European marketing business went through a significant metamorphosis, emerging with one of the most profitable and highly respected marketing operations in the European petroleum industry.

During this period, significant price wars were under way, particularly in Germany, Belgium and the U.K. Conoco fought to establish its niche as a low-price retail marketer under the Jet and Seca brands (the former Sopi stations in Germany and Austria were renamed Jet). Ultimately, the battle forced Conoco to restructure its portfolio. "We sold our Italian retail operations, as well as our underperforming stations elsewhere throughout Europe, and replaced them with better-performing, higher-throughput sites," says David Kem, president

REALIGNMENT AND REFINEMENT 173

LISA CANTU

Little did college freshman Lisa Douglas know at the time, but the young gentleman selected to be her physics lab partner would later become her husband. Lisa and Terry Cantu have worked at Conoco since 1980. For nearly 20 years, Lisa Cantu has focused on engineering drag reducers, the liquid compounds that increase flow capacity in pipelines, taking over as unit manager in 1990. Her team's research resulted in Conoco's Liquid Power Flow Improver. "We researched, developed and commercialized the product in a record two-year time span," Cantu beams. "Conoco has contributed so much to my life," she says. Cantu immigrated from South Korea to the United States at the age of 13, the adopted child of the Douglas family, and she cherishes time with her own children. "Conoco valued our family and always looked out for both of us. When I was transferred to Houston last year, Terry was also transferred into a position he wanted. How many companies go to those lengths?"

CDR 102 Flow Improver, which improves the flow properties of crude oil through pipelines, and HP Turbo 10W-30 Motor Oil were just two of many new products introduced during the 1980s.

of refining and marketing in Europe.

To leverage its production, the company entered new markets in several European countries, adding highly successful businesses in the aviation, marine and LPG markets in Ireland and the U.K. The company also introduced a Jet retail chain to Sweden in 1981, where it had bought a home heating oil company 10 years earlier. "It was a very exciting time for our marketers as they built businesses in new markets," says Kem.

New competitors entered the fray, particularly European supermarket chains, which had established low-priced gasoline retail sites at their stores to draw customers in. Conoco met them at the negotiating table. Its German subsidiary, ConMin (for Conoco Mineraloel) acquired several gasoline retailing sites from the supermarket chains and ran them as separate businesses. In Switzerland, Conoco decided on another tack, forming a joint venture with OK Coop, a large farm cooperative that had a similar supermarket-gas retail strategy.

Conoco's European retail reengineering also boasted significant innovation. In Sweden, for example, the company introduced unmanned retail gasoline stations, a concept later expanded into Denmark. In the United Kingdom, Conoco pioneered the concept of "forecourt" marketing — combining Jet retail gasoline service stations with "Jiffy" convenience stores in 1986. Many of the retail outlets of Conoco and its competitors already had small stores with limited offerings. Conoco's new concept was more than just an expansion of these stores, however, and offered the full selection and service of a small grocery. The challenge was convincing the public to change its buying habits.

"We built a 'dummy' store in the little town of Swanley to see if the concept would work," says Bill Gover, then retail sales manager in the U.K. "At the time, the thought of buying groceries from a filling station was less than appealing, so we called the stores 'Jiffy' to differentiate them from the Jet filling stations." The shops recorded extraordinary sales, and the team behind the marketing strategy received the DuPont Marketing Excellence Award in 1988.

"Jiffy sales were some 400 percent greater than sales at our other shops," Gover adds. "We led the U.K. retail industry in revenues per square foot, ahead of many High Street stores." In 1992, the *Financial Times* called Jiffy "Britain's most innovative and efficient retailing concept."

On the North American retail front, stagnant since the sell-offs of the mid-1970s, Conoco built its first wholly owned branded marketing outlet in more than a decade. The service station, which opened in Denver in the early 1980s, offered several novel self-service features, including multiproduct gasoline dispensers and electronically controlled price readouts. The convenience store at the station also had a new look — much brighter, cleaner and safer than previous outlets. It was the beginning of a new brand image in the North American marketplace for Conoco, one that would evolve through the 1980s and 1990s with several image changes. "We converted all the Denver FasGas stations to Conoco outlets to return the strength of the Conoco brand to the Denver market — our heartland," says Gary Edwards.

New products also were unveiled, such as Conoco CDR 102 Flow Improver, a chemical that improves the flow properties of crude oil to speed the oil's course through a pipeline. Introduced in 1983, CDR 102 built upon the commercial success of the company's first so-called drag reducer, CDR 101, developed by researcher Lisa Cantu a few years earlier. Both had the same objective — to provide oil companies an economical alternative to building numerous pipeline pump stations. The flow improver was made up entirely of hydrocarbons, the basic chemical compounds of crude oil, so it could be processed through a refinery just like the oil in which it was suspended.

Another new product was the Powerscrub detergent additive for unleaded gasoline, introduced in June 1986. The chemical additive "helps keep automobile fuel injectors and carburetors clean and free of power-draining deposits, giving motorists better fuel economy and reducing engine problems," *Conoco World* reported. U.S. automakers were having significant problems at the time with the formation of deposits on fuel injection systems and had asked oil refiners for help. Conoco's new, cleaner gasoline was marketed in television and print advertisements

Former Pittsburgh Steelers quarterback Terry Bradshaw was featured in several print advertisements and television commercials touting Conoco products in the 1980s. In many ads, the well-known spokesperson would urge driving consumers to "Keep Clean With Conoco Unleaded Gasolines."

MARKETING THE
CONOCO BRAND

Conoco would draw from its Western roots for many of its award-winning television commercials in the 1980s, seeking — even as Marland Oil and Continental had nearly a century before — to distinguish the company for its pioneering spirit. An old theme from the 1950s was reprised 30 years later as Conoco filmed ads for "The Hottest Brand Going." This commercial promoting Conoco's retail outlets and credit card included a gasoline oasis sprouting from the scrub-covered desert floor — actually a replica of a Conoco station built on-site. The trio of photos to the right captures actors and crew as they prepare to shoot a Conoco commercial in the still-wild American West.

REALIGNMENT AND REFINEMENT 177

Ralph Bailey is credited with keeping Conoco's spirit and integrity intact following the acquisition of the company by DuPont. The rough-hewn, plainspoken Bailey began his career as a mechanical engineer, rising through the ranks of the coal industry to become president and CEO of Consolidation Coal Company, which was acquired by Conoco in 1965. Bailey became president of Conoco in 1977 and CEO and chairman in 1979. Many Conocoans recall Bailey as extremely personable, gracious and upstanding. Among his many attributes was an uncanny ability to remember names of employees, even though he had met them only once or twice.

featuring former Pittsburgh Steelers quarterback Terry Bradshaw urging motorists to "Keep Clean With Conoco Unleaded Gasolines."

In January 1987, after twelve years as Conoco's president and eight as its chairman, Ralph Bailey announced his retirement. He had presided over the company during one of its most difficult passages, a period that stretched from the second Middle East oil crisis through numerous takeover attempts and the largest corporate merger in U.S. history at the time. Bailey's rough-hewn, bearish charm and his immensely likable personality eased Conoco through these toilsome transitions.

Although criticized by some for putting Conoco in play, Bailey fought to keep it intact and won the war. "Had Ralph not thwarted the takeover attempts by Seagram, Mobil and others, engineered the merger with DuPont and created the precedent for corporate autonomy after the merger, Conoco would have ceased to exist," says Archie Dunham. "Ralph Bailey kept this company alive."

When Constantine S. "Dino" Nicandros stepped into the spotlight to direct Conoco in 1987, the company was recognized throughout the industry as a highly efficient explorer, producer, refiner, marketer and transporter of oil and gas. But difficult market conditions — embodied by exceedingly low oil prices and much higher operating costs — confronted both Conoco and the industry. Fortunately, Nicandros, a 32-year Conoco veteran of many earlier struggles, had the experience and knowledge to plot a productive course.

Born in Egypt of Greek heritage, Nicandros had immigrated to the United States in 1955. The cargo ship on which he traveled from Port Said, Egypt, took 17 days to arrive in Boston, where he planned to attend Harvard Business School. His parents, a banker and a teacher, had given him a first-class French education and a zest for knowledge. He had finished high school at the unlikely age of 15 before entering the University of Paris, where he earned a law degree and a doctorate in economics. "I told my father that I wanted to go to Harvard next," Nicandros recalls. "He said, 'Don't you think you've had enough college?' 'No,' I told him." He was 22 years old.

After receiving his M.B.A., Nicandros reconsidered his plans to return to Egypt or Greece and instead took a job with Conoco. L. F. McCollum headed the company, which was just starting to expand overseas. With a multinational background and education, Nicandros drew the attention of both McCollum and John McLean. "Mr. Mac once said he knew Dino would succeed and someday run the company," says Cathy Wining, general manager, materials and services. "Asked how he knew this, McCollum said, 'Because he pounds the table with his intellect.'"

"Dino was much more analytical than Ralph Bailey, and given his background, very, very comfortable in the international arena," says Rob McKee. "Consequently, he moved both the upstream and downstream sectors of this company into many new areas of the world and continued to build upon our reputation as a technological leader."

Nicandros's first year in office was marked by dramatic industry changes. The production restraint shown by OPEC in the 1970s and early 1980s had loosened with a vengeance in 1986, causing a worldwide oil glut. As Arab oil supplies multiplied, prices plummeted. Meanwhile, the pressures exerted by the government's aggressive environmental agenda — a response to unrelenting bad news about air pollution, acid rain, ozone depletion and global warming — pushed costs up as prices fell. It was no surprise Conoco's 1987 earnings — $277 million — were the lowest since 1973, the year of the first Middle East oil crisis.

There were other problems. In 1982, as the political situation in Libya intensified, Conoco's U.S. citizens were evacuated quickly. "I was the last American 'expat' out," recalls John Horning, vice president of operations for Conoco's Oasis Oil Company at the time.

"I remember having to do all the packing, something my wife typically did. But she was out of the country, and the Libyan government wouldn't let her return. When I look back, I guess I never

In 1986, the Organization of Petroleum Exporting Countries (OPEC) dramatically increased production, which caused crude oil prices to plummet. Saudi Arabia's Sheik Yamani, pictured above, was the architect of the new strategy. All Western companies' profit margins were devastated. Conoco's 1987 earnings dropped to $277 million, the lowest since the first OPEC-led crisis in 1973.

REALIGNMENT AND REFINEMENT

UPHOLDING CONOCO'S
SAFETY RECORD

Throughout the latter quarter of the twentieth century, one company has consistently ranked as the safest U.S. oil company in the American Petroleum Institute's annual survey — Conoco. Shortly after he became president, Archie Dunham sent a message to all employees reemphasizing that the company's safety motto — "Our work is never so urgent or important that we cannot take time to do it safely" — applies just as much today as it did when it was adopted. Conoco has a zero-injury strategy. Employees achieved company records for fewest injuries from 1995 through 1998 and reduced injuries 35 percent during this period. In 1999, more than 70 percent of Conoco's business operations achieved this zero-injury objective.

Worldwide, Conoco employees' total recordable injury rate of 0.43 injuries per 200,000 hours on the job is three times safer than the average total recordable rate of injuries for all petroleum companies. In the industry, Conoco employees ranked number one in fewest injuries 14 out of 20 years during the span from 1978 through 1998 — a remarkable achievement. Safety pervades Conoco's daily work as well as the design of its facilities. This is true in all locations, whether in a refinery, left; or an offshore platform, above; or a training exercise, opposite.

As the industry's search for new energy sources extends into deeper water, greater depths and harsher environments, Conoco continues to develop the technologies to meet the challenges of the future. In fact, over the years, Conoco's research scientists have developed the leading seismic technology, equipment and methodologies used in the industry today. The company's seismic acquisition, processing and interpretation capabilities are at the cutting edge. At right, a Conoco scientist examines core samples for the presence of hydrocarbons. The photo opposite displays a full-color, computer-enhanced satellite image used in Conoco's exploration effort.

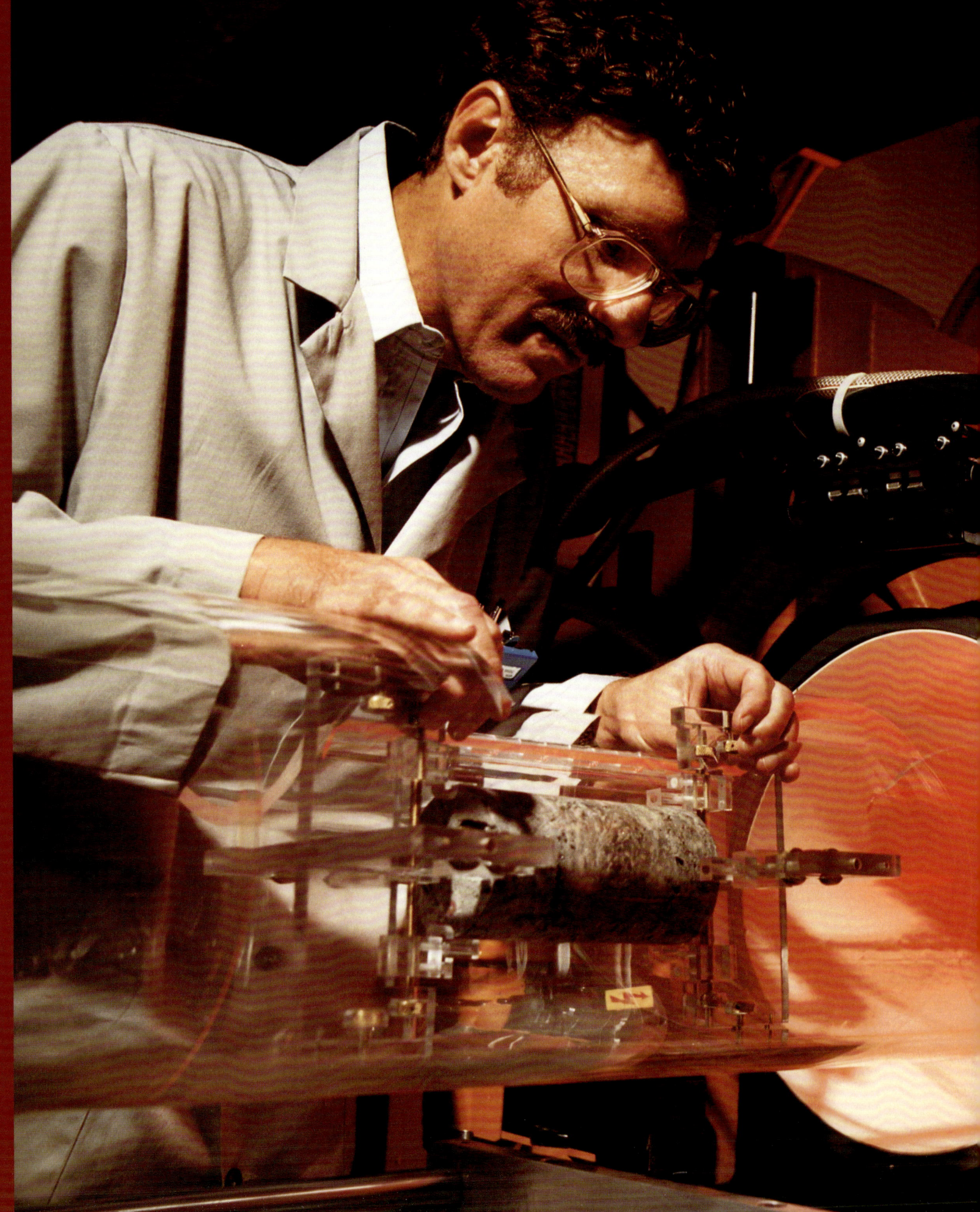

thought the pullout would last this long. I remember thinking I should find someone to sublet the house while I was gone so squatters wouldn't claim it." Ultimately, Conoco had to exit its pioneering Libyan operations in June 1986 at the insistence of the U.S. government. The company's holdings in Libya were producing about 100,000 barrels of oil a day, and the company expected this to increase over the years.

Large-scale tragedy also struck the oil industry. In July 1988, Occidental Petroleum's Piper Alpha oil rig in the North Sea exploded, taking the lives of 167 workers. The North Sea oil community is a close one, and many Conoco workers were devastated by the loss of their colleagues. In the days following the disaster, John Ogren, chairman of Conoco (U.K.) Limited, and other senior managers visited the company's manned platforms in the North Sea to calm workers and reiterate the importance of the company's safety procedures. Conoco's safety record led the U.K. government and industry to ask Rob McKee — by then, the chairman and managing director of Conoco (U.K.) Limited — to develop a public response to the tragedy.

As oil prices continued their downward spiral in early 1988, Nicandros responded with a mix of growth plans and cost reductions. To increase oil reserves, the company would expand its exploration activities and invest in new acreage. To reduce operating expenses, it would dispose of underperforming assets and squeeze costs wherever possible.

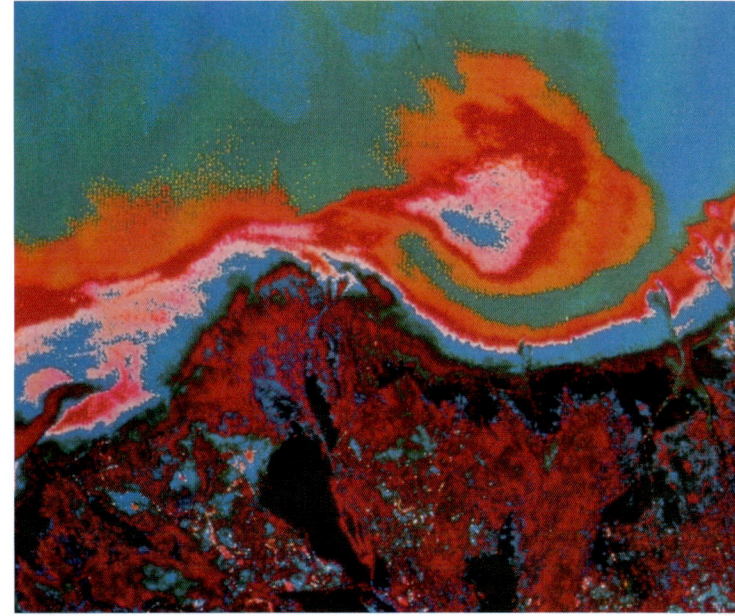

The company also would encourage entrepreneurial behavior among its employees by nurturing new ideas and rewarding the people who proposed them. "A major reason why we have always been successful in exploration, production and new technology is the *esprit de corps* of our troops and the innovative ideas they unleash," says El Grivetti, Conoco's executive vice president of exploration and production for North America at the time.

Several new upstream projects were announced, a large number of which involved natural gas — an increasingly profitable line of business. In 1969, the company formed the natural gas and gas products department, or NG&GP, to address various aspects of the natural gas business. As the years progressed, NG&GP became a gas technology center, constructed major natural gas gathering and processing systems and created one of the industry's most

CONOCO'S GENERATIONS

Many companies like to tout the family atmosphere that exists among their employees. But at Conoco, the "family" part is literal — many generations of the same family have worked for the company over the last 125 years. Indeed, many Conocoans have met their future spouses at work, only to produce the next generation of company workers! There are many multigenerational Conocoans eager to tell their inspirational and informative family histories as they relate to the company. Here are just a few.

Sue Oldfield (second from right) from Integrated Health Services in Ponca City comes from an extended Conoco family. "My grandfather, Samuel Thorpe, worked for Continental from 1928 until 1960," Sue recalls. "My father, Sidney 'Pug' Thorpe (center), began working for Continental Oil after World War II, working many years as a pipefitter until his retirement in 1985. I started at Conoco in 1996 as a nurse, and my son, William Oldfield (right), a civil engineering major, worked here one summer." Pug's brother, Jack Thorpe, and Jack's wife, Donna (far left), both retired from the tax department after 41 and 25 years, respectively, at Conoco. Inset, Donna's mother, father, aunt and uncle stand by a Conoco oil derrick in the 1930s.

A career at Conoco seemed predestined for Pat Howerton. After all, his grandfather, his father and three of his uncles all worked for the company. Pat manages the Gillis Gas Processing Plant and Louisiana Gas System. His father, Billy Joe Howerton, worked at the company from 1947 to 1989. His three uncles, Gene, Levi and Virgil, also worked for the company from the 1940s to the 1980s. His grandfather and namesake, V. C. "Pat" Howerton, inset, started work at Conoco in 1936.

Growing up, Charles David Riley (left) saw the world through his father's Conoco career. Charles William Riley started work at the company in 1968, rising through the ranks and several overseas appointments to become manager of employee relations. He retired in 1986, but his son, who goes by David, signed on in 1981 as a systems analyst. Today, he is an IM consultant in Houston.

Look up L. Walmsley at Conoco and you'll find the father-daughter team Les and Lisa. Les joined Conoco in 1969 and is currently an area maintenance leader at the Humber refinery. Daughter Lisa joined in 1998 as a financial analyst in Humber's accounts department. Lisa's loyalty to the "family" business was cemented when she received a Conoco scholarship toward her college education.

REALIGNMENT AND REFINEMENT   185

The San Juan gas-processing plant in New Mexico, which came on-line in 1986, is one of the largest and most efficient gas plants in the continental United States. The project, led by Conoco's natural gas and gas products department (NG&GP), helped motivate the renaissance of the entire San Juan Basin into one of the largest gas-producing regions in the United States.

successful management development programs.

NG&GP's commercial and technical skills were amply demonstrated by the development of the San Juan gas processing plant in New Mexico. The shining towers of this major plant are an impressive sight from the nearby highway, but the facility is only a small part of the story. Conoco had acquired the Delhi-Taylor Company in the early 1960s. With that acquisition came interests in New Mexico gas leases that were being operated on behalf of Delhi-Taylor by El Paso Natural Gas using El Paso's own gathering system and processing facility. However, the El Paso plant was an old design offering poor recovery of natural gas liquids. Conoco sought ways to squeeze more value from the Delhi-Taylor acquisition. "We chose to take over the leases and build a world-class processing plant," says Tom Knudson, a former vice president and general manager of NG&GP. "But doing so was no simple task and required commercial ingenuity."

The right to build a new facility at San Juan became part of a complex regulatory settlement that required approval by each of El Paso's several million residential and commercial gas customers in five states, from Texas to California. But under the leadership of Ted Davis, the NG&GP group found a way to get this monumental task done, and San Juan began processing natural gas in 1986. Today, it is one of the largest and most efficient gas plants in the continental U.S., and the project helped motivate the renaissance of the entire San Juan Basin into one of the largest gas-producing regions in the United States.

Other natural gas projects loomed. In July 1988, Conoco bought the Dragon Trail gas-processing plant in the Douglas Creek natural gas field in northwestern Colorado for $11 million, a project destined to process 11.3 billion net cubic feet of gas and recover 250,000 net barrels of gas liquids. Conoco's ability to produce and recover ethane for petrochemical feedstock, most notably for the DuPont plant at Orange, Texas, also increased dramatically during the late 1980s. The company became a leader in the recovery of natural gas from coal seams, with major operations in Pocahontas, Virginia, and in the "four corners" region of New Mexico and Colorado where their borders meet those of Arizona and Utah.

Conoco's natural gas ventures overseas also expanded. In the southern North Sea, Conoco (U.K.) Limited oversaw drilling and production of the "V" gas fields. The development of Victor several years earlier had opened the door to other fields in the area, namely the Vanguard, Vulcan, North Valiant and South Valiant fields. "Each of the fields had separate ownership and really wasn't economically attractive to develop on its own," says Malcolm Griffiths, director of marketing and transportation for Conoco European Gas Limited, in London. "So we figured out a way to bring all

REALIGNMENT AND REFINEMENT 187

PRIME MINISTER THATCHER
CHRISTENS THE "V" FIELDS

Prime Minister Margaret Thatcher donned a hard hat to christen the Lincolnshire Offshore Gas Gathering System (LOGGS) and the "V" fields in the North Sea. A Conocoan at the ceremony recalled: "Her security people were on high alert about a domestic security issue around that time and didn't want her to make what they saw as an unnecessary journey. Being Mrs. Thatcher, of course, she refused to be restricted and turned up at the last minute by helicopter. She immediately became deeply absorbed in everything that was going on." Accompanying Thatcher in the photo (from left) are Conocoans John Ogren, Dino Nicandros and Roger Ramshaw.

British Prime Minister Margaret Thatcher inaugurated the four "V" fields in the Southern Basin of the North Sea and the Lincolnshire Offshore Gas Gathering System (LOGGS). The inauguration culminated the largest natural gas project in the U.K. in more than a decade. In what was fast becoming a Conoco tradition, the project was completed ahead of schedule and under budget at a total cost of $1 billion. Production in the fields is highly automated, with unmanned wellhead platforms linked by subsea flowlines to a central gas-gathering station and network. LOGGS provides economical transportation for natural gas from the four original "V" fields — Vulcan, Vanguard, North Valiant and South Valiant — to the gas terminal at Mablethorpe in Lincolnshire. Many other fields and producers in the region currently use the LOGGS facilities.

13 owners together to develop one pipeline and pool the gas. The contractual challenges far outweighed the logistical ones, but we did it."

The pipeline — called the Lincolnshire Offshore Gas Gathering System, or LOGGS — transports the gas to the Viking terminal at Theddlethorpe, where it is processed, metered and dispersed. LOGGS, together with the original Viking Transportation System and the Caister Murdoch System, transports some two billion cubic feet of gas per day — roughly 20 percent of the United Kingdom's daily gas requirement. The four-year project, christened by Prime Minister Thatcher in September 1988, was completed ahead of schedule and below the original project estimate of $1.3 billion.

Another gas project was undertaken in Indonesia by Conoco Indonesia Inc. in a partnership between Conoco and Pertamina, Indonesia's state-owned oil company. In January 1990, the Alu Alu E-1 well north of Jakarta and east of Malaysia in the West Natuna Sea struck both gas and oil. The well flowed at a daily rate of 12,289 barrels of crude and 58.8 million cubic feet of natural gas. The new field was named Belida — like Alu Alu, an Indonesian name for a fish. In subsequent years, the Belida field would firmly anchor Conoco's major upstream presence in this region of the world.

Conoco made its first foray into the Caribbean with its entry into Trinidad in 1989. That year, Conoco struck a deal to construct a gas plant in Point Lisas to supply feedstock for neighboring industrial customers and natural gas liquids for export to the Caribbean and the Americas. The enterprise, dubbed Phoenix Park Gas Processors, Limited, was a partnership between the National Gas Company of Trinidad and Tobago, Conoco, and Pan West Engineers and Constructors.

"Shell originally had a lock on the project," says Rick Oshlo, who was general manager of international gas at that time. "Pan West, with whom we had worked in the San Juan Basin in the southwestern U.S., alerted us to the deal. When it went to an international bidding round, we offered the government a much better equity position and were able to put together a plan that was significantly cheaper, more effective and efficient. Shell was working hard at lobbying high government officials, which they're good at, but we decided instead to work with the National Gas Company and bring them to our side of the table. Ultimately, they carried more influence with the government, and we got the project."

Operations for the $100 million modern gas-processing facility began in 1991 with a capacity of 650 million cubic feet per day. Since then, the gas plant has been expanded twice, evolving into one of the largest and most sophisticated gas-processing facilities in the Western Hemisphere and a significant contributor to Conoco's natural gas and gas products earnings. With a capacity in 1999 of 1.35 billion cubic feet of natural gas per day, the plant was able to meet a significant portion of the natural gas liquids demand throughout the Caribbean and northern South America.

At a news conference on April 10, 1990, a few days before the 20th anniversary of Earth Day, Conoco announced the most far-reaching environmental initiatives ever undertaken by an oil company. Nine initiatives were announced in all, but the one that received the most press attention was the company's resolve to use double-hulled tankers to transport oil and to commit to their use exclusively.

The decision was a mix of altruism and expediency. The previous year, the Exxon *Valdez*, a single-hulled tanker, foundered in Alaska's Prince William Sound, spilling 11 million gallons of oil. Double-hulled ships would reduce the likelihood of such spills in the future but were much more expensive to build and buy. Consequently, the American Petroleum Institute, the oil industry's trade association, fought attempts in Congress to require their use.

Conoco broke ranks with the industry, becoming the first oil company to commit to building only double-hulled crude oil ships. "It took Nicandros five minutes to decide to build the first tanker, the *Patriot*, and we stuck with double-hulled vessels," says Antonio Valdes, head of the company's

DOUBLE-HULLED TANKERS

Conoco received worldwide attention in 1990 when then-CEO Dino Nicandros announced the company's initiative to use only double-hulled oil tankers like the model above. Conoco was the first oil company in the world to make such a commitment, and both the public and the press applauded the move. Hundreds of editorials were written about the initiative, and the company received a flood of supportive letters written by everyone from schoolchildren to politicians. The environmentally progressive step helped restore a measure of public faith just a year after the Exxon *Valdez*, a single-hulled tanker, had run aground in Alaska's Prince William Sound, spilling 11 million gallons of oil.

The *Guardian*, the second double-hulled tanker to be built for Conoco, proved the efficacy of the protective double-hull design. In 1997, the ship sustained a 100-foot by 4-foot gash in its outer hull, seen above right, as a result of a collision with a tug and barge flotilla near Lake Charles, Louisiana. Despite the severity of the accident, not a drop of oil was spilled. Opposite, workers at the Samsung Heavy Industries shipyard in Korea discuss the construction of one of Conoco's double-hulled vessels, also shown below.

worldwide marine group, who suggested the policy. Another tanker was ordered in 1990, the *Guardian*, and two more in 1991, the *Pioneer* and the *Continental*.

The decision received enormous press attention. The day following the announcement, an article with Nicandros's picture appeared on the front page of the *New York Times*, and editorials in more than 100 newspapers praised the company's actions. Nicandros personally received hundreds of congratulatory letters — from business and political leaders to ordinary Americans, who scribbled words of thanks on their Conoco credit card bills. "It was a shot in the arm — for Conoco and for me personally," he says.

The company's objective of operating 100 percent double-hulled fleets in U.S. waters by the year 2000 was achieved two years earlier in 1998. The fleet consists of four Aframax tankers and 14 tank barges in U.S. waters, plus a double-hulled tanker that shuttles crude oil in the North Sea. Two new double-hulled tankers added in 1999 completed another Conoco milestone of operating an all double-hulled fleet in international waters. The North Sea shuttle tanker — originally the *Heidrun*, now renamed the *Randgrid* — first proved the wisdom of this policy in 1996 when it struck an uncharted rock off the western coast of France. The rock ripped large holes in the bottom of the tanker. Thanks to the double hull, however, not a drop of the one million barrels of oil in cargo was spilled, as the inner hull remained unscathed.

A year later, Conoco's prudence again paid off. The *Guardian*, one of Conoco's tankers operating in U.S. waters, was rammed by another vessel, causing a 400-square-foot rip in its outer hull. "Since the tanker was fully loaded, this could have been an environmental nightmare," says Dennis Parker, vice president of safety, health and the environment. "But not one drop of the ship's 550,000 barrels of crude oil escaped into the Calcasieu River at the Port of Lake Charles, Louisiana."

The decision to employ only double-hulled ships extended to two new deepwater drillships commissioned by Conoco and R&B Falcon Corporation, ships that would explore some of the deepest waters of the world.

The remaining environmental initiatives of 1990 included pledges to develop cleaner fuels, construct only double-containment systems at all new retail stations, reduce toxic air emissions and hazardous solid wastes beyond existing legal requirements and create Citizen Advisory Councils to help the company implement local environmental programs. Nicandros continued to speak out forcefully on the environment. He made Conoco's position crystal clear: "We must consider the environmental consequences of everything we do."

For the first time in many years, the public had something to cheer about in an oil company. "Dino deserves credit for giving Conoco a kinder, softer focus as a company, especially at a time when the oil industry received so much hostile press and public reaction," says Archie Dunham. "He confronted the industry with initiatives others would not dare, taking the steps he knew were right for society and for future generations."

Nicandros also released a vision statement at Conoco's annual management meeting in May 1991. He stated what Conoco is, "a premier integrated, international petroleum company"; its role, "to find and develop oil and natural gas, to manufacture a wide range of petroleum products, and to transport and market them through a variety of channels"; and how it will accomplish its objectives, "in a superior way, and by doing so, serve our customers very well."

The speech touched upon the four core values important to Nicandros — safety, ethics, the environment and people. "I emphasized that to me, ethics were more important than profit, safety was critical, diverse individual backgrounds were respected and needed by the company, and we were all a part of society and must serve it responsibly," Nicandros says today. "I recall it being a very impassioned speech."

As the summer of 1991 drew to a close, Nicandros again would be called to express his emotions. On September 4, a company airplane carrying 12 members of the Conoco family crashed

HONORING
LOST LEADERS

The news that 12 members of the Conoco family had died in a plane crash sent emotional shock waves through the company. A Conoco corporate jet headed to the Middle East and Asia-Pacific regions crashed in the mountains of Borneo, killing the passengers and crew. They were husbands and wives, parents, friends, leaders, visionaries — and their loss was deeply felt. The tragedy gripped employees and the local press as the community absorbed the devastating news. The blow to the company was monumental, but even greater was the personal sense of grief that Conocoans felt. Tributes were held around the world, and in Houston, more than 3,000 people attended a memorial service honoring Kent and Connie Bowden, Bill and Gayle Dietrich, Colin and Brooke Lee, Jim and Linda Myers, Ann Parsons, Ken Fox, Steve James and Gary Johnston. "This group leaves a legacy across the entire company and the globe," Dino Nicandros said. "Each one leaves his or her own mark that is stamped upon our memories for the rest of our lives."

REALIGNMENT AND REFINEMENT   193

in Southeast Asia while approaching the airport at Kota Kinabalu on the island of Borneo, killing all those aboard. The nine passengers and three crew members had set off from Houston three days earlier on a business trip that was to include visits with government officials and joint-venture partners in the Middle and Far East. Among the dead were several high-ranking Conoco executives and their spouses.

At a memorial service on September 12, attended by more than 3,000 people at Houston's Second Baptist Church, Nicandros paid tribute to those who had died: "Our hearts are heavy. Our grief is deep. Our loss is great. Our friends are gone." His moving eulogy was simultaneously broadcast to hundreds of Conocoans around the country and in the U.K.

"These were people I worked alongside for many years on so many projects," Nicandros says, looking back. "They were the best in every way, the cream of the crop. So their loss was both a personal and an institutional tragedy."

"It took a long time for the company to recover," says Archie Dunham. "These were people we loved, people we respected. We knew each other's families, each other's kids. Despite all the difficulties this company has endured over its history, I'd have to say the plane crash was our lowest point, on an emotional level. All of us in the oil business travel extensively to many distant parts of the world. The crash made everyone view their lives, their work and their families with a much different perspective."

Many key management changes were effected in the wake of the tragedy. "These people had worked hard to see Conoco succeed," Nicandros says. "Now it was up to us not to disappoint them."

Discouraging as it was as far as corporate earnings were concerned, 1992 was a year of considerable achievement in the areas of exploration and production. Conoco's earnings of $337 million were below expectations, a consequence largely of crude oil prices, which had declined 7 percent from 1991 levels. The company's oil reserves, however, increased significantly. Three new fields were brought on-stream in Indonesia and the North Sea, including the Miller field, about 160 miles (258 kilometers) east of Aberdeen, Scotland, which added about 34,000 barrels of oil and 58 million cubic feet of natural gas to daily production volumes.

While these projects offered hope for the future, the oil price crash continued to devastate the industry. Aside from a brief spike up during the 1990 Persian Gulf War, prices and margins were falling year after year. It was painfully apparent that a crude price rebound was increasingly remote, "an ever-receding bonanza," a retired executive lamented. Survival would depend on being able to make money in a market where prices are always soft.

"Our only hope, I told my team, was to get costs

In 1992, Conoco published its Project 2000 Blueprint for U.S. upstream and, in 1993, its Global Excellence Blueprint for international upstream. The goal of these plans was a more focused business achieved by an intense restructuring effort. By the middle of 1993, the company had disposed of underperforming domestic assets and reduced its international operations, retaining assets in only 10 countries (down from 19 in 1991). Four core upstream regions were identified — North America, the North Sea, Dubai and Indonesia. The restructuring helped the company's upstream earnings per barrel jump 14 percent from 1991 to 1994.

way down," says Nicandros. "'Either we do it,' I told them, 'or we're gone.'"

Thus began the most ambitious and agonizing corporate restructuring in Conoco's history. The company benchmarked itself against competitors and divided into internal teams that worked with an outside company, Gemini Consulting, to instill "best practices" in every corporate discipline and process, from treasury to travel to technology. Assets were rearranged, jobs were eliminated, and the previous function-based organizational structure built by Howard Blauvelt was overhauled. "The Gemini study gave us the courage to change from a functional organization, with one vice president for refining, another for marketing and another for transportation, into a business unit value chain that integrates these various functions," explains Jim Nokes, Conoco's vice president of refining and marketing for North America.

These initiatives spread to every facet of the company. In 1992, Conoco published its Project 2000 Blueprint for U.S. upstream and, in 1993, its Global Excellence Blueprint for international upstream. At the same time, it unveiled a downstream reengineering effort called Target 2000.

Heading the six-member management team directed to draw up the Global Excellence Blueprint were Conoco's executive vice presidents for exploration and production — Archie Dunham, who oversaw North America, Russia and Europe, and Rob McKee, responsible for Africa, the Middle East, Asia-Pacific and South America. The programs analyzed Conoco's global upstream business in great detail and developed courses of action.

The answers seemed to be the same — Conoco's upstream performance did not compare well with its competitors. The company's management spent much of 1992 and early 1993 defining the difficult choices necessary to improve. It was a challenging time for the newly formed Upstream Leadership Team and its two senior leaders. McKee brought tremendous domestic and international upstream experience to the task. Dunham brought the management expertise from NG&GP, downstream and two major DuPont businesses, one of which, Polymers, had undergone a major restructuring under his leadership.

Noted industry analyst Mike Mayer, of Wertheim Schroder & Co., added his evaluation to the process. It was bad news. Mayer intensively analyzed 15 companies to compare upstream performances. Based on 1992 data, Conoco was number 15. Dunham and McKee convened the upstream management group and reviewed this devastating study. "We simply had too many people, too much management and a poorly performing portfolio of properties," Dunham recalls. McKee remembers the stark reality that "we had to rationalize our asset base, fundamentally change our portfolio and drastically reduce our costs." Difficult restructuring and portfolio decisions continued well into 1994. It was a tough period, but results began to improve almost immediately.

Says Nicandros, "We forced everyone here to look at their operations to identify those activities we should build up and those we should cease. We pulled back on exploration in several areas and sold or otherwise disposed of several hundred fields that were not yielding proper returns."

At the end of 1991, Conoco held interests in 476 North American producing fields. By the end of 1993, the company had concentrated its interests into fewer than half this number while keeping its production volume at almost the same level. In addition, it cut the number of countries where Conoco had upstream operations from 19 in 1991 to 10 in 1993. Four core upstream regions were identified: North America, the North Sea, Dubai and Indonesia. "By concentrating our operations, we were able to reduce the number and the costs of the regional offices needed to support these field operations," Nicandros says.

"Moreover, the experience we gained over time exploring, developing and producing within one area improved our working knowledge of things unique to that area — both geological and cultural — enabling us to operate more efficiently."

By the mid-1990s, the strategy was reaping dividends. Upstream earnings per barrel jumped 14 percent from 1991 to 1994, despite a $4 drop

Conoco streamlined and reorganized its North American downstream assets in 1993 into three regional business units — the Gulf Coast, Mid-Continent and Rocky Mountain regions. Conoco's retail network throughout the world was rationalized, from more than 7,000 stations to about 5,000. In 1988, Conoco and Kayo Oil merged their marketing operations, and Kayo began selling gasoline under the Jet name in many parts of the U.S. Eventually, Conoco initiated a campaign to link the company's foremost North American brands in the minds of customers. Jet stations benefited from the connection to Conoco's quality image, and the Conoco brand gained strength from increased exposure at more outlets across the nation.

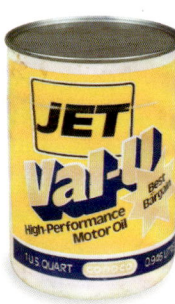

The emergence of free enterprise in Eastern and Central Europe presented Conoco new opportunities to reach consumers there. The company began a multiyear expansion program in 1991, opening six Jet outlets in the former East Germany. Above is the Potsdam, Germany, station with a Conoco office in the building behind.

in crude prices during the period. Even with the rationalization of producing assets, the company's oil and gas reserves reached an all-time high — 1.85 billion barrels of oil in 1995. "We would not have achieved this growth had we not refocused our exploration programs," Dunham says today. The changes begun in the early 1990s paid a satisfying dividend in 1999 when Mike Mayer issued a new report based on 1998 statistics which placed Conoco first in upstream performance when compared to 15 competitors.

Conoco's downstream structure in the United States also was streamlined and reorganized in 1993 into three regional business units — the Gulf Coast, Mid-Continent and Rocky Mountain regions. "We wanted to align downstream's organizational structure more closely to our assets and customers," Nicandros says. As part of a rationalization of Conoco's retail network throughout the world, the number of stations was reduced from more than 7,000 to about 5,000.

"I'd say 80 percent of our ultimate cost reductions resulted from changing the way we managed our assets," says Nicandros. "The rest were reductions in manpower." The number of employees had already been pared to about 22,000 by 1992, half the employment count of 1975. Most of this reduction related to the separating of Consol in 1987 and Conoco's chemical business several years earlier, but Conoco was forced to make additional job cuts through the decade.

"Mike Rocconi, vice president of human resources, led Conoco's HR function with courage and effectiveness through the downsizing of the 1990s," says Archie Dunham. "He championed the interests of our employees and pushed the development of innovative programs — like Hire Conoco — which allowed us to retain, and reassign, many hundreds, perhaps thousands, of valuable Conoco employees who otherwise would have been outplaced."

By 1996, these cost reductions and business improvements would add some $700 million to Conoco's 1995 earnings — a remarkable achievement. "Without the improvements, our 1995 earnings would have been essentially break-even," says Rob McKee. "It just shows you how much change was necessary to survive and be competitive in the new market environment."

As the company rationalized its asset base in the United States, it expanded its downstream presence in several new markets, including Norway, Spain, the Czech Republic, Poland, Hungary and Thailand. Dramatic changes in Eastern and Central Europe presented Conoco new opportunities to reach consumers in these newly emerging markets. The company began with a multiyear expansion program in 1991 by opening six Jet outlets in the former East Germany. "I attended the ribbon cutting for the first retail outlet in Rostock," recalls Gary Edwards.

"I was amazed to see people lined up for blocks in their cars waiting to buy our gas. Then, after filling up, they waited in another line to go to our car wash. Through an interpreter, I asked a customer why she would wait in line so long just to buy gas. She replied, 'We've waited 45 years. What's an hour?'"

Conoco's more targeted exploration approach hit bull's eyes, for instance in Nigeria, where test wells confirmed significant oil deposits in the Niger delta, leading to the discovery of the Ukpokiti field. An even more promising new region was the former Soviet Union, heretofore unavailable to Western oil explorers and producers.

The Cold War had thawed following Mikhail Gorbachev's declaration of *glasnost* and *perestroika* in 1989, and Soviet authorities signaled an openness to foreign investment. Max Pitcher, executive vice president for worldwide exploration, assembled a small team and traveled extensively in the Soviet Union in 1990 to assess the prospects. As a business relationship began to develop, Dino Nicandros visited with future Russian President Boris Yeltsin in Moscow in 1991, just a few months before the fall of the Soviet Union. Meetings held during this trip solidified early relations with the reformists who would lead the new Russian Federation. Nicandros says, "Max Pitcher was truly the one responsible for putting us on the map in Russia. Its enormous future potential impressed us all."

Conoco ventured where no other company dared, forging the first joint venture between American and Russian companies to develop a new oil field. The venture, Polar Lights, was formed in 1992 after many years of relationship-building and negotiations. In 1991, Dino Nicandros met future Russian Federation President Boris Yeltsin, left, during a negotiating trip to the Soviet Union.

REALIGNMENT AND REFINEMENT 199

The unique ecology of the Arctic tundra demanded creative solutions from the hundreds of Conocoans and Russians who developed the Ardalin field. They built roads out of ice to transport equipment and supplies to the remote site in the winter months. When the ground thawed, helicopters were needed to lift and lower equipment directly into place. Raised drill pads and an above-ground pipeline — complete with reindeer crossings — were built to protect the nomadic way of life on the delicate tundra. By 1996, Ardalin's production had reached 25,000 barrels a day and despite constant political turmoil, increased to 40,000 barrels a day in 1999.

"I had a 'hunting license' to go there and look for opportunities for Conoco," recalls Pitcher. "I met with Vladimir Semenovich, an old friend from the 1970s who was then chief geologist for oil exploration in the Soviet Union. He told me there was an area north of Moscow on the Arctic Sea where oil had been discovered but not developed. I remember he outlined the geology of the region on a napkin in my hotel room."

The area targeted for exploration was the Timan Pechora Basin west of the Ural Mountains above the Arctic Circle. Reserve potential was estimated at one billion to two billion barrels of oil, with the first field to be developed — Ardalin — having recoverable reserves estimated at 110 million barrels. Pitcher set his sights on developing the region, but as Russia was a new country finding its way, he had difficulty determining who was in charge. "Russia was an enigma wrapped in an enigma," he says, paraphrasing Winston Churchill. "I talked with the minister of geology who said, 'The state owns the minerals. I am the state. You will make your deal with me.' Unfortunately, he was later dismissed after he sided with the wrong group during an attempted coup."

Although the negotiations were delicate, Russia strongly wanted to develop the Ardalin field and needed Conoco's help. "The Arctic coast is some of the most sensitive terrain on the planet," Pitcher says. "The country's environmental activists were very concerned over oil field development. What appealed to them about us was our ethic of environmental stewardship and safety. They welcomed our best practices."

In June 1992, the negotiations were concluded, and a deal was struck. Conoco forged the first joint venture between a U.S. oil company and a Russian concern to develop a new field. Named Polar Lights, the 50-50 venture between Conoco International Petroleum Company, a Conoco subsidiary, and Arkhangelskgeologia, a Russian geological association, was modest by Conoco standards, but it set the stage for future endeavors. "Our investment of about $375 million over three years

Dunham, above, went to unusual lengths to secure the buy-in of local Russian communities, such as the Nenets, even sampling raw reindeer meat, a local delicacy!

REALIGNMENT AND REFINEMENT 201

MICHAEL W. BRITTON

Mike Britton was 20 years old before he saw life beyond the border of Sturgis, Michigan, but after 26 years with Conoco, he says, "I've had the opportunity to travel the world several times over and live and work in many exciting places. In the process, I've learned the entire upstream business, working as a researcher, engineer, economist, executive assistant and manager." He cites Ardalin in Russia as among the most personally rewarding projects he worked on. "We accomplished several firsts with Ardalin, most notably in the environmental area," says Britton, currently project manager for Petrozuata in Venezuela. "It is very easy for me to see the impact of my contributions. That gives me a great deal of pride and enthusiasm. Back in 1973, I had offers from several oil companies. I knew Conoco would give me the freedom and opportunity to make the most out of my career and myself. I've never regretted my decision."

was not that much in an overall context, since we were investing about $2 billion all over the world at the time," says Nicandros. "But we wanted to get a foothold in Russia. Ultimately, we gained valuable operating experience within Russia's changing political and economic environment."

The hundreds of Conocoans and Russians who worked to develop Ardalin installed drilling rigs, built storage tanks and constructed an elevated, above-the-ground pipeline to avoid damaging the permafrost — in order to feed oil south into the Russian fuel system. They worked around the clock, mainly in winter, because the Russian tundra was too spongy to move heavy gear in the summer months without causing environmental degradation. "It was a bog," says John Horning, then president of Conoco International Petroleum Company. "One of the first things we had to do was build an ice road to haul the rigs and equipment. And we used heavy-lift Russian helicopters to fly in all our camp equipment — including the living quarters. The pilots were amazing; they'd place these 12-ton modules within a foot of the support stations they were to reside on. It was something."

The Polar Lights project was closely followed in the business press. "We were the first foreign company to develop an oil field in Russia from scratch," Horning notes. "But our investment was shrouded by political uncertainty. Skepticism ran in equal measure with optimism."

In August 1994, after two and one-half years of development, the project started pumping oil. By 1996, Ardalin's production had reached 25,000 barrels a day, and several years later, it was producing 40,000 barrels a day. "For those who doubted the feasibility of investment in Russia, the Polar Lights project is graphic proof of the viability of such investments," Russian Fuel and Energy Minister Yuri Shafranik said at the time.

While Nigeria and Russia offered hope for the future, Conoco still viewed the North Sea as its primary production area. In the spring of 1991, Conoco Norway Inc. received permission from Norway's parliament to develop the giant Heidrun oil and natural gas field in the Norwegian Sea, about 70 miles (113 kilometers) south of the Arctic Circle. Discovered by Conoco in 1985, Heidrun was a tremendous find. It was the northernmost offshore oil field to be developed and one of the world's largest discoveries at the time. It had estimated recoverable oil reserves of 750 million barrels and 1.6 trillion cubic feet of natural gas — a giant field by any standard.

To develop the field, which lay below some 1,150 feet (351 meters) of water, the company's engineers built the world's first concrete tension-leg platform, which represented another generation of the TLP concept Buck Curtis and his engineers had pioneered a decade before. Using lightweight aggregate concrete reduced the weight of the hull

Everything about Norway's Heidrun field merits the word giant. Discovered in 1985 near the Arctic Circle in the hostile Norwegian Sea, Heidrun ranked as one of the world's largest new discoveries during the 1980s. The total project cost of $4 billion made it Conoco's biggest project ever. The pioneering TLP (pictured above being towed to sea) is an industry giant

The necessary size of the Heidrun platform and the depth at which it would be moored — some 1,150 feet (351 meters) of water — demanded a new generation of Conoco's TLP concept. The company's engineers responded with the world's first concrete tension-leg platform. By using light-weight aggregate concrete, the weight of the hull at sea was reduced by as much as 40 percent from that of conventional concrete, which increased its buoyancy — even at a weight of 220,000 tons — and hence its deck-load capacity. The massive concrete support beams of the platform and deck were floated from the onshore construction site to the assembly site in a Norwegian fjord, right.

at sea by as much as 40 percent from that of conventional concrete, increased buoyancy and also provided more deck-load capacity.

"Heidrun was a giant field, but it posed some challenges that were unique at the time," says John Kemp, who managed Conoco's Norway operations at the beginning of the Heidrun project. "The challenge of deepwater gave us the chance to prove just how good the tension-leg platform concept is."

The $2.8 billion platform was massive. It weighed 220,000 tons in all, with a deck the size of an American football field and a platform 12 times the height of the Statue of Liberty. The platform included facilities for more than 350 workers, although only 250 workers would be on-site during normal production. The platform's life span was estimated at 50 years, and the platform itself made up approximately 80 percent of the cost of the multibillion-dollar project.

Because Heidrun was in a remote area far from the existing natural gas infrastructure, Conoco Norway had to develop a viable plan for using the field's gas reserves before the government of Norway would approve the project. A decision was reached with Statoil, Norway's state-owned oil company, to pipe the gas to Tjeldbergodden in mid-Norway, where a world-class methanol plant would be built to convert the feedstock. The plant, which ultimately cost $1 billion to build, drew heavily on DuPont's methanol production expertise.

Conoco owns 18.125 percent of the Heidrun field, as well as the methanol plant and the pipeline. Altogether, Heidrun was developed for a cost of $4.1 billion, the company's largest, most expensive project to date and one of the costliest projects ever undertaken by any oil company. But given the potential oil and gas reserves and the technological knowledge that would be gained, it was well worth it. "We pushed the envelope — engineering, logistical, political — in every part of the project," said Mike Johnson, president of Conoco Norway at the time.

Heidrun was christened in May 1995. It currently produces about 240,000 barrels of oil a day. Looking up at the giant structure during the dedication ceremonies, Nicandros says, he was completely awestruck. "It just took my breath away."

When he announced his retirement, Nicandros was hailed for transforming Conoco into a highly competitive international petroleum company, despite the most difficult market conditions in the industry's history. "The fact that he was able to increase the company's value during a time of falling oil prices and market instability is a tribute to both his leadership and his mastery of global economics," says his successor, Archie Dunham.

As the sun set on 1995, the helm of Conoco would be passed to a new navigator. Dino Nicandros, a pillar of strength during the difficult transitions of the late 1980s and early 1990s, retired that year.

Indeed, before he handed over the helm in January 1996, Nicandros had lifted Conoco's revenues to a record $17.7 billion in 1995, while paring more than $700 million from its overhead and operating expenditures. He had firmly established Conoco as the industry's leader on environmental issues and had added luster to its technological prestige. Moreover, Nicandros had opened the door to Russia and Eastern Europe, had guided the company into ever-deeper waters in the North Sea and the Gulf of Mexico and had paved the way for its future large-scale ventures in Venezuela and Malaysia.

Asked to look back on his 39-year career at Conoco, Nicandros referred to his initial interview with Conoco, reflecting that he might not have had any relationship with the company at all "were it not for the huge bowl of shrimp the company's recruiters offered me back in my days at Harvard" — a hilarious comment coming from such a refined individual. Then he paused and uttered but one word — "exhilarating."

At Conoco's annual Holiday Concert in Houston in 1995, the company's employees presented Dino Nicandros with a ship's bell. The symbolism was clear — he had steered the company skillfully through some very rough seas. Nicandros would not live to see Conoco reach its 125th anniversary. In August 1999, after battling illness for several months, he passed away.

Known for his refinement and razor-sharp intellect, Constantine S. "Dino" Nicandros brought to Conoco a new kind of leadership that enhanced the company's international stature. Born in Egypt to Greek parents, Nicandros was educated in Paris, France, before immigrating to the U.S. to complete an M.B.A. at Harvard. He spoke several languages fluently and maintained a lifelong interest in the arts. Following the merger with DuPont in 1981, Nicandros effectively managed Conoco's effort to gain post-merger alignment. He subsequently managed all of Conoco's petroleum operations before being named chairman, president and CEO in 1987. Nicandros led Conoco during challenging times. His accomplishments include the company's many international projects during the period, major successful corporate restructurings and industry-leading environmental initiatives. To commemorate his retirement in 1995, Nicandros was presented with a ship's bell, recognizing his skill in navigating the company through rough seas.

He was born and reared in the oil patch in the state E. W. Marland made famous for oil — Oklahoma. He also shared the great oilman's passions for geology, education and, above all, people. But, where Ernest Whitworth Marland ended his career with a company lost, Archie Wallace Dunham would distinguish his with a company born anew.

Small-town virtues gave Archie Dunham the values by which he has conducted his life — a love of God and family, a respect for others and a belief in honest, hard work. His mother was raised on a farm and sewed colorful shirts for her son from feed sacks, clothing he was proud to wear. His father was an oil field gauger for the Pure Oil Company in Ada, Oklahoma. "He'd get up at 4 a.m. and drive to the oil field to gauge the oil tanks and start the pumps that would transport the oil to Chicago," Dunham recalls. "Then we'd go squirrel hunting."

Dunham loved rocks as a kid, a lifelong hobby that would lead him into the field of geology. He became the first in his family to attend college, earning a degree in geological engineering from the University of Oklahoma. Following a four-year

stint with the U.S. Marine Corps in Hawaii, Dunham and his wife, childhood sweetheart Linda Burns, returned to Oklahoma, where Dunham earned a master's degree in business administration, also at the University of Oklahoma. He interviewed with 18 companies after graduation and was offered positions at 17. "I was about to join IBM when I got a phone call from Bob Lee, a Conoco manager, who was on campus and wanted to have dinner," Dunham recalls. "I was making about 80 cents an hour wrapping books at the university's press, and Linda was a secretary in the College of Education. We had two kids and no money. I heard 'dinner' and said 'sure.'"

Lee told Dunham about Conoco's management development program, an intensely competitive and arduous executive training regimen in which only the fittest survived. "Bob said if 10 graduates entered the program, three might last the year, if they were lucky," Dunham recalls. "He told me the workload would be unbelievable. I'd have to absorb a tremendous amount of knowledge about the industry, technology and finance in a short time. Then, every few months, I'd be grilled by five or six executives with about one hundred years of experience among them. It would be terrible."

So, in 1966, Dunham signed on.

In his first 10 years with the company, he held various operating and commercial positions, mostly in natural gas and gas products, ascending the corporate ladder. In 1976, he was sent to California as the executive vice president of Douglas Oil, becoming its president three years later. He moved back to Houston in early 1981 as vice president of logistics and downstream planning. In 1987, Ed Woolard, DuPont's president and CEO, asked Dunham to become the first Conoco executive to accept a position at DuPont as its senior vice president of chemicals and pigments. While at DuPont, Dunham later led the polymers sector, the second largest operating segment of DuPont. When Dino Nicandros retired

On October 22, 1998, 30 percent of Conoco's stock began trading on the New York Stock Exchange. The IPO was greatly oversubscribed, and at $4.4 billion, it was the largest in U.S. history at that time. It was Archie Dunham — seen here with his wife, Linda — who put in motion plans for the separation of Conoco from DuPont. The son of an Oklahoma oilman, Dunham had enrolled in Conoco's management development program in 1966. He led Conoco's Douglas Oil and later DuPont's polymers sector before succeeding Nicandros as Conoco's president and CEO in 1996. In 1998, Dunham argued his rationale for two strong, independent companies to DuPont's board of directors. They agreed a reborn Conoco would benefit both companies and their shareholders. After 17 years as a DuPont subsidiary, Conoco was on the road to independence.

INDEPENDENCE 211

Personal accountability was the cornerstone of a new Conoco. Employees would be challenged to perform to higher corporate expectations, for which they would receive incentive compensation. The program was called Conoco Challenge.

as Conoco's president and CEO in January 1996, no one was more prepared to succeed him.

The new CEO had high hopes for the company. In a letter to employees, he wrote: "I want nothing less than for Conoco to be recognized around the world as a truly great, integrated, international oil company that gets to the future first." He listed what this journey would require — focus, leadership, balance, character and flexibility — the "five basic attributes of great companies," he said.

He invited employees to join him in this quest. "I'm here to share with you my expectations, to listen to your concerns and ideas, and to respond to them," Dunham wrote. "People can't be expected to trust or be inspired by their leaders if they don't see or get to know them."

Dunham is completely at ease in all company. He talks with visitors not from behind his desk, but eye-to-eye in a pair of comfortable chairs. "He has that unpretentious Oklahoma ability to appeal to everyone," says Liz Williamson, executive assistant to the president, chairman and CEO. "You look him square in the face and realize, 'I like this guy.'" Indeed, Dunham's e-mail box is typically crammed with messages from employees. Says Williamson, "He always writes back."

"Archie believes in giving employees plenty of room to take risks, to challenge decisions and even to make mistakes," says Tom Henkel, vice president of investor relations. "It's all part of what he believes is the need for people to have the freedom to dream. His job, he often says, is to provide them with the resources to help them do their best."

In March 1996, Dunham pronounced a bold objective for Conoco — to double its value to $30 billion within eight years. To achieve this ambitious goal would require an earnings increase of at least 10 percent per year — no easy task in an industry beset by low pricing. But Dunham insisted the strategy was necessary "if we are to become the outstanding company I want us to be."

As a critical first step, Dunham unveiled a sweeping restructuring, a transformation covering virtually every facet of Conoco and every individual within. Job roles and responsibilities were revamped, business units were given more flexibility, a companywide training organization was created, and in-depth strategic business reviews were instituted. Eight hundred employees were moved off corporate rolls in such areas as finance, human resources and legal affairs, and into Conoco's business units. In place of the former corporate staff, the Conoco Leadership Center was created, comprising about 150 individuals. Dunham called the restructuring what it essentially was: "Fundamental Change."

Personal accountability would be the cornerstone of this new Conoco. Says Dunham, "I made a pact with employees that if they perform up to their own and my expectations, and Conoco achieves success because of it, they will be rewarded." In 1997, the second year in a row in which Conoco reaped record earnings, the new CEO came through on the "Conoco Challenge," a program Nicandros had spawned in 1995. More than 11,000 nonmanagerial employees worldwide received a total of $48 million in special bonuses. "Receiving a Conoco Challenge award makes me feel as though I have contributed to the success of the company," says Randol Marzuola, project manager for the carbon fibers group. "It makes the 'thank-you' very tangible. In meetings, the phrase, 'that affects my Challenge,' is often said. The undercurrent of that is that any decision we make affects the company's profitability and therefore our potential for individual reward. It personalizes every company decision."

Not since Dan Moran's unexpected largesse at the close of the Great Depression had Conocoans shared so greatly in the company's prosperity. "Even people outside the company are amazed that we are rewarded in such a manner," says Juanita Garner, volunteer coordinator. "This makes me feel like Conoco is really doing something different. Conoco Challenge sends a really strong message to me and other employees that Conoco recognizes us and the contributions we make to the company. And the bonus always seems to come right on time!"

"For me, Conoco Challenge raised my personal awareness of the financial goals of the company," says Caroline Bratton, gas marketing and communications. "It clearly defined how I, as an individual

ANDREW ROBERTS

For Andrew Roberts, living in Houston has brought to light more than occasional language barriers and culture shock. "Conocoans from different countries and disciplines all seem to possess certain down-to-earth qualities of openness and willingness to knuckle down to the task at hand," says Roberts, who joined Conoco (U.K.) Limited in his native London in October 1987 and now leads market development initiatives in the carbon fibers business. "Recently, we've developed a technology that makes our carbon fibers available in larger quantities and at lower cost than in the past, opening up a vista of new markets for us — such as concrete, asphalt and composites — in which carbon-fiber-reinforced resins or pitches would replace steel or aluminum as a fundamental raw material," he says. "This is a revolutionary concept, the result of a corporate philosophy that allows local management to control its own destiny with relatively unobtrusive central intervention," he explains. "Now, if we could all learn the same language.... Just kidding!"

THE CONOCO
CHALLENGE

When Conoco reaped record earnings in 1997, Archie Dunham delivered the Conoco Challenge, rewarding a total of $48 million in special bonuses to more than 11,000 non-managerial employees. Below, Dunham cuts the "pie" celebrating corporate achievement of the 1997 Challenge goal, themed "Your Piece of the Pie." At the festivities in Houston, Dunham told employees, "We're here to celebrate sharing this big pie with the people who made it possible." The CEO joined other executives in serving real apple pie to the enthusiastic throng. In other Conoco locations, workers symbolically shared in the pie they had earned, eating pizza pie in the Rockies and Australian meat pie in Brisbane. In Stavanger, Norway, kransekake pie was served, and in Jakarta, boxed Indonesian-style sweets. In February 1999, "coach" Dunham was doused with confetti, left, as he and fellow employees held a "pep rally" to celebrate reaching the 1998 Challenge goal.

contributor, have a role and a responsibility, through how I do my work, to influence not only my specific business unit's success or failure, but ultimately the corporation's overall profitability as well."

In the mid-1990s, the U.S. oil industry was beleaguered by declining growth in established markets, such as North America and Western Europe. An oversupply of oil, stagnant demand and cutthroat pricing would take their toll, compelling several oil companies to merge and consolidate later in the decade.

In addition to worldwide market conditions, U.S.-based petroleum companies, including Conoco, faced another obstacle — U.S. foreign policy.

Early in the decade, Conoco began negotiations to develop two oil fields off Iran's Sirri Island in the Persian Gulf and to export the associated gas and NGLs to nearby Dubai. Conoco's Middle East team, led by Brian Neylon, underwent three years of intensive negotiations. In early 1995, the National Iranian Oil Company (NIOC) signaled to Conoco that they were ready to formalize agreements. A full team, including Dino Nicandros and Archie Dunham, converged on Tehran. Final negotiations took place during a grueling day and night at NIOC's palatial guesthouse in the northern suburbs of Tehran.

"Dino and I would walk in the gardens of the

guesthouse with our other team members so we could confidentially discuss the remaining areas of dispute," Dunham recalls. "The hours wore on, and tensions were evident on both sides. We finally sealed the deal around midnight and raced to the airport believing that NIOC and Conoco had created a history-making agreement. Euphoria marked our long flight out of Tehran to Dubai. The following day, we arrived in Europe and heard President Clinton's criticism of the agreement. Conoco had done everything correctly. We had violated no U.S. law and had kept the U.S. government fully informed throughout the whole process. We were deeply discouraged."

Conoco's team immediately went to Washington for consultations with the U.S. government. Mike Stinson, in his role as vice president of new business development, testified before the U.S. Congress, but Washington only confirmed their worst fears. President Clinton issued two Executive Orders canceling Conoco's project and prohibiting American companies from investing in Iran or purchasing Iranian crude oil. After almost four years of negotiations, Conoco's landmark Sirri project was canceled only a week after it was signed.

However, President Clinton's unilateral sanctions against Iran held little sway with U.S. allies. Oil companies in France, Malaysia and Russia moved in to exploit the opening crafted by Conoco. Eventually, France's Total went on to develop the Sirri fields.

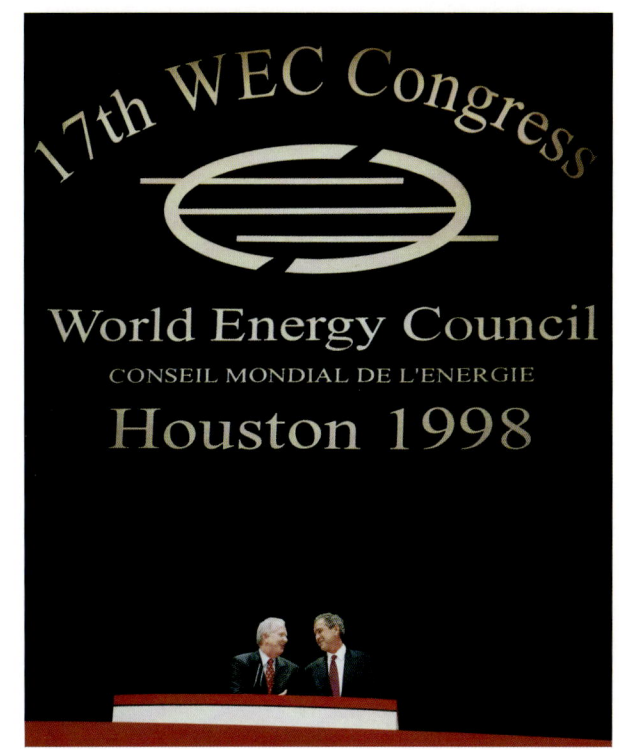

Archie Dunham and Texas Governor George W. Bush, a candidate for U.S. president in 2000, share the dais at the 17th World Energy Council (WEC) meeting held in Houston in September 1998. There, Dunham spoke on global energy issues, having stepped firmly into the spotlight of energy policy in 1995 when the U.S. government issued new unilateral sanctions against the Iranians. Conoco lost its pioneering Sirri oil and gas project in Iran, and Dunham became a leading critic of unilateral economic sanctions and a sought-after speaker on the topic.

Dunham seized on the issue of sanctions, writing numerous opinion articles and eventually becoming a sought-after speaker on the subject of economic sanctions. "I don't question the goals of economic sanctions," he wrote in a *Washington Times* piece. "But if sanctions are unavoidable, we should try everything in our power to make them multilateral. It is joint pressure from many countries that is most likely to bring meaningful results. Unilateral sanctions only result in American companies losing out to companies from other nations."

"Sometimes it seems that sanctions have become virtually the only tool in our government's foreign policy kit," he stated in one speech. "When I see the list of potential target countries, I'm reminded of the old saying: 'When your only tool is a hammer, every problem starts to look like a nail.' Well, I'm inclined to wonder," he continued, "if we are hitting our thumbs more often, and harder, than the nail."

As the decade progressed, U.S. policy makers began to take note. In March 1999, the U.S. government lifted the ban on shipments of food and medicine to Iran, Sudan and Libya. In addition, Congress drafted legislation that would require the government to consider the cost and likely effectiveness of economic sanctions before imposing them. In July 1999, the U.K., a staunch ally of the U.S. in the implementation of economic sanctions against Libya, resumed diplomatic relations with the

INDEPENDENCE 215

Demand for fuel in Eastern and Central Europe surged in the 1990s, as this queue of drivers at a Jet station in Rostock, Germany, demonstrates. Conoco expanded into newly opened markets in Poland, the Czech Republic, Hungary, Turkey and Slovakia. By 1999, nearly 3,000 retail stations in 17 European countries offered the company's products.

Libyan government, and hopes rose among the U.S. petroleum industry that reforms in U.S. policy would be forthcoming.

Earlier in the decade, several markets formerly closed to foreign investment had opened their doors. With a reputation for operating efficiency, innovation and proven project management ability, Conoco was an appealing candidate for countries seeking outside partners.

The company's explorers and producers found a revived interest in the Asia-Pacific area. Geoscientists David Bond and Janice Christ developed an innovative theory indicating the promise of large reserves in the deep waters off the coast of Vietnam. By the early 1990s, Vietnam was opening to American participation, and Conoco's commercial and scientific team landed major new deals with the Vietnamese. "It took over four years of difficult negotiations to gain their trust and confidence," notes Dee Simpson, business development manager. Bill Lafferrandre, who served as Conoco's manager in Hanoi, adds, "But now Conoco has positioned itself to become the leading American company in Vietnam." Oil and gas exploration will begin in 2000.

Jack Blackshear and Sing Wong produced a similar success in securing two very large exploration blocks off the coast of Cambodia. The company also explored most of the continental shelf in the Straits of Taiwan through an innovative joint venture with Chinese Petroleum Co. of Taiwan negotiated by Conoco's Daisy Liu.

Elsewhere in the world, demand for fuel in Eastern and Central Europe was surging, and Conoco quickly set in motion plans to expand its solid downstream base in Western Europe to include these newly opened markets. It had already established Jet outlets in the former East Germany in 1991. As Poland, the Czech Republic and Hungary embraced free market enterprise in the 1990s, Conoco made modest investments in those countries as well, opening 17 outlets altogether by 1993. As the decade progressed, the company added Turkey and Slovakia to its network. By 1999, nearly 3,000 retail stations in 17 European countries offered the company's products.

To supply these burgeoning markets, Conoco sought proximate refining capacity. In September 1995, after two years of negotiations, the Czech Republic government gave the green light to Conoco and two partners, Agip and Shell, to purchase a 49 percent interest in refineries at Kralupy and Litvinov. The two refineries operate as a single entity with intermediate streams moving between them. "The deal marked the first major refining privatization in Central and Eastern Europe, and the first with Western partners," says Dave Kem. The Czech refineries began drawing from Conoco's oil production in Russia and other sources, including the North Sea. Conoco plans to integrate North Sea gas into the Czech system by feeding it into a new cogeneration plant supplying both electricity and steam to the refinery.

As downstream gained a foothold in Eastern Europe, Conoco had an eye on the Asia-Pacific region as well. Demand for petroleum there was rising 5 to 7 percent annually in the early part of the decade, four times the rate in the United States and the United Kingdom. In 1994, Conoco signed a historic joint-venture agreement with two partners — Petronas, the national oil company of Malaysia,

In 1994, Conoco partnered with Petronas, the national oil company of Malaysia, and Statoil, with whom Conoco had worked in other parts of the world, in a historic joint venture. They signed an agreement, right, to meet growing demand for refined petroleum products in the Asia-Pacific region by constructing a new refinery near the Malaysian city of Melaka, above.

INDEPENDENCE 217

and Statoil, with whom Conoco had already worked in other parts of the world. The partners constructed a refinery near the city of Melaka to produce a full range of refined petroleum products to help meet the growing demand for motor fuels, lubricants and specialty products in the region.

Conoco's proprietary delayed coking technology is being used at the Melaka refinery — the first time this world-class technology has been employed by Conoco in a new refinery in the Asia-Pacific region. "Petronas had been looking for a way to build a refinery to process heavy, sour crude but had been discouraged by the high capital costs," says Paul Lashbrooke, who retired in September 1999 as vice president of refining and marketing, Asia-Pacific. Conoco had the bottom-of-barrel refining technology Petronas needed. "We showed them they could reduce their capital requirements by using our delayed coking technology, which turns low-value residual oils into higher-value gasoline and diesel oils," Lashbrooke adds.

Conoco has a 40 percent interest in the 100,000-barrel-a-day refinery, which came on-stream in 1998. Crude oil from the company's Indonesian and Dubai production facilities, as well as from other locations, is refined at Melaka into a variety of products, including gasoline, fuel-grade coke, naphtha, LPG (liquefied petroleum gas) and distillates. The refining project ultimately cost $1.4 billion, with Conoco chipping in roughly

The 100,000-barrel-a-day Melaka refinery, left and opposite, came on-stream in 1998. Crude oil from the company's Indonesian and Dubai production facilities is refined at Melaka into a variety of products, including gasoline, fuel-grade coke, naphtha, LPG (liquefied petroleum gas) and distillates. Conoco's proprietary delayed coking technology is also in use at the Melaka refinery. The prized technology converts low-value residual oils into higher-value gasoline and diesel oils. The refining project ultimately cost $1.4 billion, with Conoco chipping in roughly $600 million. In the next millennium, demand for energy is expected to grow rapidly in the Asia-Pacific region, accounting for a large percentage of Conoco's downstream value.

INDEPENDENCE 219

S VICRACHAT

S Vicrachat is a mechanical engineer with a passion for design. Born and reared in Bangkok, a city he describes as the most beautiful in the world, Vicrachat (his given name) has led the design team building Conoco's retail outlets in Thailand over the last seven years. "I design the interiors on my own and work with others to design the exteriors," says Vicrachat. He sits down with the company's co-branded "food partners," a well-known group including A&W, KFC and Dairy Queen, to create an original design. "Some of them look very Western, others are more Eastern-looking, and still others are a mix of the two," the Conoco project engineer says. Vicrachat handles more than just design, construction and co-branding efforts. "I am also responsible for maintaining all the stores in our network, as far as any repairs and maintenance," he says. "Ask anyone in Thailand who has the cleanest stores and they will say Conoco."

$600 million — its largest-ever investment at the time in a single downstream facility. "It was well worth it," Lashbrooke says. Gary Edwards agrees: "Despite the recent economic downturn, Asia-Pacific will still account for a significant percentage of downstream's value in the longer term, given the steady demand for fuel in the region."

On the retail side, the company's growing marketing operations in Thailand featured more than 100 retail stations in 1999. Besides Thailand, Conoco is examining other nearby markets for Melaka's output, including Indonesia, southern China and eastern India, where the company opened a business development office in Bangalore in 1996.

Conoco also became the first foreign company in 30 years to obtain a license to develop Malaysian retail outlets selling gasoline, other fuels and also grocery products. "The Malaysian government said we had to use a Malaysian or 'Bumiputra' brand name unique to that country," Edwards says. "So we created ProJET to play off our well-known Jet brand."

By 1999, Conoco either owned or had a significant stake in four refineries in Europe — two in the Czech Republic, one in the United Kingdom

Melaka helped supply Conoco's burgeoning marketing operations in Thailand and other Asia-Pacific markets, including Indonesia, southern China and eastern India. Conoco also became the first foreign company in 30 years to obtain permission to sell gasoline in Malaysia. Opposite, Conoco Executive Vice President Gary Edwards bows before Buddhist monks at the opening ceremonies for the first Jet station in Thailand in 1993. By 1999, there were more than 100 Jet retail stations, like the one above, in Thailand.

INDEPENDENCE 221

IVAN McCONKEY

In the small town of Grimsby, England, Ivan McConkey is restoring the 1915-era house he and his wife bought not long ago. The house fits McConkey's view of himself as "someone who tries to make things run better." McConkey joined Conoco in 1980 upon receiving his chemical engineering degree and today heads the Humber refinery's coke optimization team, which recently spearheaded the commercial development of a new grade of coke, Heat Soak Super-Premium. The new coke performs better in the end-use application — electric arc furnaces — and offers improved economics. Had it not been for Conoco's emphasis on personal accountability and teamwork, McConkey says, the new coke would still be a dream. "The company sets the goals and then leaves it up to managers and team leaders to achieve what are often far-reaching objectives. If the objectives are easy, you become a medium performer. When they're tough, it brings out a team's ingenuity and innovation."

and one in Germany — to serve its marketing operations in Austria, Belgium, the Czech Republic, Denmark, Finland, France, Germany, Hungary, Luxembourg, the Netherlands, Norway, Poland, Slovakia, Spain, Sweden, Switzerland, Turkey and the U.K. "The key to our downstream success has always been a retail presence centered on a highly efficient refinery," Kem says. "When you hear our competitors talk about our enviable situation in Southeast Asia, Western Europe or Eastern Europe, what they're really talking about is our refining position."

During the decade, Conoco's Humber refinery was upgraded to improve its production of needle coke and anode-grade petroleum coke — raw materials sold to produce electrodes and anodes used in the global steel and aluminum industries. "In the typical refinery, low-valued fuel coke is a by-product of the heavy crude refining process," says Tom Mueller, Humber refinery manager. "We do things the other way around and actually are driven by premium-valued coke." The refinery's expertise in these specialized product lines earned it four Queen's Awards for Export Achievement.

In the U.S., Conoco undertook an ambitious refinery project to produce another specialty product. In April 1994, Conoco announced the Lake Charles refinery would be the focus of a $500 million joint venture with Pennzoil, a major marketer of lubricating oils. The project, called Excel

Conoco's Lake Charles, Louisiana, refinery, seen here at night, is the site of a $500 million joint venture with Pennzoil called Excel Paralubes. The two partners built a hydrocracker to produce high-quality base oils from low-quality crude. The high-quality base oils produced at Lake Charles are the primary ingredient of more than 300 different finished lube products for commercial and industrial use, among them Conoco's new class of lubricants, Hydroclear. The project makes Louisiana the hydrocracked base oil capital of the world.

Paralubes, has earned Lake Charles an industrywide reputation as a refinery able to make a silk purse from a sow's ear. "It takes the worst-quality crude in the world and makes the highest-quality petroleum-based lubricants out of it," says Jim Nokes. "That's what I call a powerful upstream-to-downstream business chain."

The high-quality base oils produced at Lake Charles are formulated into more than 300 different finished lube products for commercial and industrial use, among them Conoco's new Hydroclear line of lubricants. Other finished products are anticipated. "We take these hydrocracked base oils, which are

Hydroclear is less expensive than synthetics, outperforms solvent-refined lubricants and makes equipment last longer. It raised the standard of heavy-duty oils. Here, Conoco scientist Gerald Graham, tests the clarity and viscosity of Hydroclear oil samples at the company's Ponca City research laboratory. The finished product goes into Conoco's valuable line of Hydroclear oil products, right.

INDEPENDENCE 225

In the halcyon days of the oil business, a "gasser," described a drilling endeavor that produced natural gas and not crude oil. Given rising global demand for natural gas, a gasser today is cause for celebration. "Natural gas will take a larger share of the energy pie because it has become the fuel of choice in electricity generation due to its low environmental footprint and attractive economics," says Marianne Kah, chief economist and manager of business and market analysis. To capitalize on expectations that the demand for gas will grow at a higher rate than the demand for oil, Conoco has funded natural gas exploration and development projects in the North Sea, Indonesia, Norway and the United States, positioning the company to become a major player in the burgeoning natural gas industry. "Conoco is focusing on areas for gas development where we have a competitive advantage in terms of geological knowledge or infrastructure," says Kah.

Among the most important natural gas projects of the decade was Britannia, a giant gas and condensate field in the central North Sea. Britannia's reserves are so extensive, they alone could meet the total domestic and industrial gas demand of the United Kingdom for at least a year. With estimated reserves of about 3 trillion cubic feet of gas and an expected life span of 30 years, the field offers peak production of 740 million cubic feet of gas and more than 50,000 barrels of condensate per day — enough to power 1.1 million

In 1989, Conoco struck a deal to construct a gas plant in Point Lisas, Trinidad, to supply neighboring industrial customers and export LPGs to the Caribbean and the Americas. The enterprise, Phoenix Park Gas Processors, Limited, is a partnership between the National Gas Company of Trinidad and Tobago, Conoco and Pan West Engineers and Constructors.

high in purity, low in aromatics and look as clear as water, and process them further into so-called 'white oils,' which are safe for human consumption," says John Derr, general manager for Conoco's lubricants and specialty products business unit. "White oils are used in a great many pharmaceutical and cosmetic products. It's really a specialty business with a lot of promise."

Conoco had reached an agreement with Chevron in 1991 to operate Britannia jointly, a novel arrangement. In addition, there were about a dozen different oil companies involved in developing the project, requiring enormous diplomacy. "To keep the project participants working together and to hold costs down, we put together delicately negotiated risk-sharing alliance contracts," says Jeff Tetlow, the project director for Britannia.

As the project moved forward, Rob McKee made calculated moves to increase Conoco's share of the field, purchasing several interests and exchanging some oil holdings in the North Sea for additional pieces of Britannia. An agreement between Conoco and Oryx traded Conoco interests in the Murchison, Hutton, Lyell and other fields for Oryx's 15.5 percent interest in Britannia. Ultimately, Conoco's total interest in the giant gas field increased from 8 percent to 43 percent.

In early 1995, just a few weeks after the exchange with Oryx, the partners received the go-ahead from the U.K. Department of Trade and Industry (DTI) to develop the field. In announcing the decision, U.K. Minister for Industry and Energy Tim Eggar called Britannia "one of the most important offshore gas field developments ever undertaken."

"We used to think that the biggest challenge on major projects was the technology, or the logistics or some other 'hard' issue," says Tetlow, who also led the design and construction of the $1.5 billion Britannia facilities for Conoco. "On Britannia, we learned that the critical element is motivating everyone — employees, contractors, suppliers — to jointly achieve an extraordinary result."

The motivating factor for the members of the team was a greater degree of inclusiveness in the development of the project and more autonomous control of their respective capital budgets. Team members participated in an induction program that explained the project's vision and values. "We created a contractual framework that was nonconfrontational and removed the restraints under which contractors normally worked," Tetlow explains. The result was a steady stream of creative solutions. "Ultimately, the unique alliance contracts alone helped cut about $450 million from the original project cost estimate," says Malcolm Griffiths.

When the field came on-stream in August 1998, it was at a total cost of $2.05 billion, 20 percent below the original forecast. Most importantly, the Britannia team built and installed the eight-legged, 48,400-ton production platform ahead of schedule while maintaining Conoco's industry-leading safety and environmental performance standards.

This culture of environmental accountability and safety is deeply imbedded in Conoco's activities and is as important to the success of a project as the budget and schedule. Employees take personal responsibility for Conoco's safety and environmental performance, while managers and supervisors,

U.K. homes when operating at full capacity.

Bringing the project on-stream would prove formidable, but Conoco brought to the table a record of achievement in project management. Two separate independent studies, in 1992 and 1994, concluded that Conoco's projects not only cost less but were completed faster than the industry average. "We do the right projects and the projects right," company literature states. Britannia would put this maxim to the test.

The development of the Britannia field in the U.K. North Sea was one of the most important natural gas projects during the 1990s. At a peak production of 740 million cubic feet of gas and more than 50,000 barrels of gas condensate per day, the field can power 1.1 million U.K. homes. Conoco reached a novel agreement with Chevron in 1991 to operate Britannia jointly. As the project moved forward, Conoco made calculated moves to increase Conoco's share of the field from 8 percent to 43 percent. The Britannia team built and installed the eight-legged, 48,400-ton production platform ahead of schedule and under budget while maintaining industry-leading safety and environmental performance standards. These photos depict the giant platform at sea and one worker making the daily rounds.

The goal of achieving an extraordinary result motivated the teams of Conoco employees, contractors and suppliers building Britannia. Team members cited a greater degree of inclusiveness in the development of the project and more autonomous control of their respective capital budgets for the ultimate success of the project. For example, drilling teams pictured here achieved a number of records, including the longest hole of its type drilled at that time.

Pictured is an employee at the Conoco-operated Viking gas terminal at Mablethorpe on the Lincolnshire coast. The terminal processes natural gas delivered from several Southern Basin gas fields and transfers it to the British gas grid. England's narrow streets prove a formidable challenge for this process vessel, opposite, as it is being delivered to its new home.

guided by formal plans and processes, are committed to making these the top priorities in all their operations and activities.

The development of the Britannia project was a proving ground for the practicability of these long-held objectives. "Looking back, Britannia was the pinnacle, as far as our project management skills are concerned," McKee says. "It's the very best job ever done in this industry. We did it for less money than forecast, were on-stream earlier than expected and put it together via a delicate, complex contractual alliance that broke new ground. At present, we're producing more from this field than we ever expected."

Dunham agrees: "Before Britannia, Conoco and Chevron were fierce competitors seeking to outdo each other at all costs, as opposed to finding ways to work together to benefit both companies. Our ability to successfully manage this project together marks a new direction for the industry. Moreover, the reputation we earned as a skilled project manager has helped us secure creative, cost-competitive financing for several other world-class projects."

During the decade, production from Conoco's many gas activities found its way to market. From the Norwegian Sea, gas from the company's Heidrun field was transported to its newly completed, world-class methanol plant at Tjeldbergodden, Norway. In April 1998, the first gas from the small Boulton field in the southern North Sea, which the

Pictured are two views of the giant Troll platform in the North Sea. The field was discovered in 1979. Gas production started in 1994 at a rate of 800 million cubic feet per day, with a peak rate of 1,500 million cubic feet per day reached in 1996. The field is expected to last another 15 years. Conoco holds a 2 percent interest in Troll, adding significantly to its European portfolio, which includes Britannia, the original Viking and Victor gas system, the "V" gas fields and the Lincolnshire Offshore Gas Gathering System (LOGGS), the Caister-Murdoch System (CMS) and the Statfjord gas system.

company had developed in 1997, arrived at Conoco's Theddlethorpe, England, natural gas complex — three months ahead of schedule. By building a less costly unmanned platform rather than expensive subsea facilities, Conoco saved a third of the customary project costs, making the otherwise marginal Boulton field economical to develop.

Britannia, of course, was the capstone of Conoco's significant gas assets in Europe, which included the original Viking and Victor gas system, the "V" gas fields and the Lincolnshire Offshore Gas Gathering System (LOGGS), the Caister-Murdoch System (CMS), the Statfjord gas system and the giant Troll gas field in the North Sea. Driven by the promise of mammoth future production from Britannia, in 1994 the company acquired a 10 percent interest in the new Interconnector gas pipeline in the U.K. The $700 million pipeline, completed in 1999, connects the U.K. to Europe at Zeebrugge, Belgium, improving Conoco's ability to supply major gas markets throughout northwest Europe, as well as the U.K. market. "For the first time, we can sell natural gas that we produce in England to Germany, France and beyond," says Don Robertson, general manager, extraction, in

INDEPENDENCE 233

AVIS A. BRAGGS

While everyone else is celebrating the new millennium, Avis Braggs is at Conoco headquarters in Houston, "one of many on our information management staff standing vigil," she says. Braggs, a native of Trinidad and the coordinator of the Conoco Global Year 2000 Program, interfaced with every company facility on a global basis and assisted Conoco's many external partners to develop Y2K readiness. "Where we have disruptions on a normal basis, we've proven to have excellent contingency plans in place," Braggs says. She served on the American Petroleum Institute's Y2K task force for two years, working with other industry leaders to share information and address concerns. "It gave me tremendous insight into how we compare to other companies," she says. "What I learned — and I'm not stretching this — is that we tend to do things better," she admits. "And our people are given more freedom and authority to get things done. Compared to others I met, I felt very empowered to do what I needed to get done."

Aberdeen. As George Watkins, chairman and managing director of Conoco U.K., said during Britannia's official inauguration in December 1998, "The Interconnector puts Britannia at the heart of Europe."

Europe was not the only region targeted for the company's burgeoning natural gas enterprises. Southeast Asia emerged as a viable market for gas, and Conoco set its sights — and technological expertise — on developing another natural gas project, this time in Indonesian waters.

The company had discovered significant oil and gas deposits in Indonesia's West Natuna Sea in the 1970s, but the remote location and lack of a commercial market suppressed early hopes of economically developing the gas. "We had acreage holdings off Indonesia and had discovered some significant oil deposits, particularly in the Belida field," says Ted Davis. "Along the way, when we were drilling for oil in the Natuna Sea, we found something else — a lot of gas. Unfortunately, there just wasn't a big enough market then to justify development. Nevertheless, in the back of our minds, we always knew that someday we'd find a way to commercialize the finds."

In early 1996, a Jakarta-based team led by Project Manager Tom Bundy, working with Conoco business developers Mary Ann Pearce and Les Porter, brought a new approach to the stalled effort. They developed a plan to make the reserves commercial, confirmed that Singapore was the strongest market and then began very complex negotiations to turn the plan into reality.

A combination of efficient new technologies and increased market demand in Singapore enabled Conoco to make a breakthrough in July 1997. "Our project teams — and there were several over such a long period — just kept hammering away," says McKee. "A few times they got to the altar, so to speak, only to be jilted. Finally, they broke through — by convincing the Indonesians to export and the Singaporeans to buy."

After more than two years of talks, Pertamina, the Indonesian state oil company, agreed in July 1997 to supply Singapore with 2.5 trillion cubic feet of gas, nearly half of which would come from nine Conoco-operated gas fields in the West Natuna Sea. The 25-year deal was finalized in 1999 between Pertamina and Sembawang Engineering and Construction, providing for the first international sale of natural gas from offshore Indonesia to Singapore, via a 300-mile (483-kilometer) subsea pipeline. The project involves building a pipeline and a major gas production, treatment and compression system to be completed in order to facilitate the delivery of gas by mid-2001.

"Singapore has a secure supply of gas to power its planned industrial development over the next 20 to 30 years," says Jean "Pogo" Davis, then manager of international gas and gas products.

Workers install the foundation for the Belida platform in the West Natuna Sea. The name Belida comes from an Indonesian word for a type of fish. The field was discovered by partners Conoco and Pertamina, Indonesia's state-owned oil company. Though Conoco had discovered both gas and oil in Indonesia's West Natuna Sea decades earlier, there was no market for the gas at the time. In 1997, after two years of talks, Conoco convinced Pertamina to begin export of natural gas to Singapore, below, via a 300-mile (483-kilometer) subsea pipeline.

INDEPENDENCE 235

In July 1997, Conoco completed a $929 million acquisition of gas reserves and assets in South Texas's Lobo trend, its largest acquisition in 30 years. The deal, which included 1,100 miles (1,770 kilometers) of gas-gathering and transportation pipeline and 215,000 acres of leases, made Conoco the second-largest producer in the state of Texas and serves as the cornerstone of the company's North American gas business. Lobo, the Spanish word for wolf, symbolizes Conoco's agile recovery methods in the region. With the benefit of Conoco's expertise in 3-D seismic imaging, drillers, right, are experiencing 90 percent success rates from the Lobo wells.

"And Indonesia, after a period of turmoil, has a vote of confidence in its future from international oil companies and the government of Singapore, as well as a concrete new revenue base."

And very importantly for Conoco, the deal created the foundation for building a natural gas business in Southeast Asia. "The project will provide long-term cash flow and a competitive advantage in marketing additional gas volumes as we add reserves through exploration in the Natuna Sea," McKee says. "Right now, this production comes from Indonesia, which is the foundation of our Southeast Asia upstream plans. But in the future, as we continue to explore in Vietnam, Cambodia, Thailand and elsewhere in the region, the entire Asia-Pacific region will become more important to us. It's why we have to be there."

The third major international market for Conoco's impressive natural gas enterprises is in North America. The company has undertaken one of the largest natural gas ventures in its history in a region of the U.S. where it has deep roots — South Texas. "We don't want to be known as an oil company that does all its investing overseas," McKee says. "We have always believed that the United States would remain an important part of our history."

Growth through acquisition would become the cornerstone of Conoco's U.S. natural gas business. In July 1997, the company completed a $929 million acquisition of gas reserves and assets in South Texas's

Lobo trend. The deal included 1,100 miles (1,770 kilometers) of gas-gathering and transportation pipeline and 215,000 acres of leases. Since the acquisition, production from the acquired properties increased from 510 million cubic feet per day at year-end 1997 to 750 million cubic feet per day at year-end 1998. Conoco earmarked $1 billion to increase net production in Lobo between 1997 and 2002 to roughly one billion cubic feet of gas per day.

"This acquisition — our biggest in the last 30 years — literally transformed our presence in the Texas natural gas business," says Ted Davis. "We went from being the twelfth-largest producer in the state to being the second largest."

Conoco's natural gas assets from the Lobo trend will be supplemented with plans to lay 100 miles (160 kilometers) of pipeline each year between 1997 and 2002 to accommodate the needs of South Texas shippers. Conoco earmarked $1 billion to step up its drilling program, left, over the same time period and increase net production in Lobo. Above are Lobo workers, who met their goal of zero injuries while keeping up with the demands of pipeline construction. On the facing page, a Lobo derrick is framed against the setting South Texas sun.

And Conoco is still increasing this North American presence. "Several large basins in the U.S. offer great potential, including the Gulf of Mexico, Texas and New Mexico," McKee says. "We're targeting each of these and developing a sizable foothold. These projects frequently offer quick payback. At Lobo, for example, our expertise allows us to drill wells rapidly, put in pipelines quickly and then quickly produce the reserves."

At the heart of this speed is Conoco's work with 3-D seismic imaging. Company geoscientists use this technology to create maps of drilling targets offering virtually pinpoint accuracy — precise enough to increase Conoco's drilling success rate from 20 percent to 90 percent. "Lobo is the Spanish word for wolf, a nimble, independent animal," says McKee.

"And that's exactly how we try to run our business."

It would have been impossible to predict that natural gas — a hydrocarbon that petroleum companies once routinely burned off — would move Conoco forward so quickly in the 1990s. On an average day in 1999, Conoco's many natural gas ventures around the globe combined to produce more than 1.3 billion cubic feet of gas and 100,000 barrels of natural gas liquids, not to mention the 500 million cubic feet of natural gas it purchased for resale and the additional 1 billion cubic feet it traded. Altogether, Conoco handles nearly 3 billion cubic feet per day or 1 trillion cubic feet of natural gas per year. These are numbers that would have astonished the early oil hunters, but they bode well for Conoco's future. Says Mike Stinson, "It appears that in many markets, but particularly in North America and Europe, natural gas may eclipse crude oil as Conoco's major upstream product."

It has long been known in the oil industry that deep inside Venezuela is a treasure trove of crude oil rivaling the grandest deposits in the world. But Venezuela was closed to foreign participation and development, keeping outside oil companies on the buying end of this coveted supply. All this changed in the mid-1990s, when Conoco broke through a gauntlet of barriers to secure a significant presence in the region.

"In late 1990, Venezuela announced that within the decade, it would open the door to foreign investment in its oil sector," says Dave Griffith, a business development manager. "We sent a couple of delegations of our senior executives to visit their senior executives. They talked over a long list of possibilities, but it didn't come to anything."

Griffith was a Conoco oil trader at the time, in

Breaking through barriers — political, financial and technological — is an apt description of Conoco's historic agreement to develop Venezuela's vast Orinoco Belt. Petrozuata — the joint venture with Petroleos de Venezuela, Venezuela's state-owned oil concern — took six years of intense negotiations to structure. When finally inked in 1995, the deal represented the first international joint venture in Venezuela after 20 years of nationalization. Petrozuata is expected to produce up to 2 billion barrels of extra-heavy crude between 2000 and 2035, boosting Conoco's reserve base. A 125-mile (201-kilometer) pipeline was constructed to transport the oil from Zuata to an industrial complex near the city of Jose on the northern coast. Pictured here is a rare sight of pipe sections before the line was completed and buried.

In August 1998, the first barrel of diluted extra-heavy crude entered the pipeline to Jose. The inauguration of the project included the blessing of the first oil by a Catholic priest. The $1 billion upgrader facility at Jose is based on Conoco's delayed coking technology, which will be used to upgrade the heavy crude into lighter synthetic crude and various by-products. This proprietary technology was a critical part of Conoco's appeal as a foreign partner for the development of the Orinoco Belt.

COMMUNITY DEVELOPMENT
IN VENEZUELA

Agroforestal, a wholly owned subsidiary of Petrozuata, employs workers at a tree plantation in eastern Venezuela. There they raise native species, such as eucalyptus, a fast-growing plant species harvested for the manufacture of oriened strand board (OSB), a type of aggregate wood material used in the construction industry. The trees from the plantation are used to offset wood pulp imports to Venezuela. The tree-planting program is part of a larger effort — a community development plan supported by Petrozuata to help ensure a sustainable future for the area's people. Not only does the project provide much-needed jobs and critical raw materials, it promotes the "greening" of the area, exemplifying Conoco's commitment to the environment.

charge of supplying crude oil to the Lake Charles refinery. In July 1991, he led a team to Caracas to meet with Jorge Zemella, the head of new ventures at Petroleos de Venezuela (PDVSA), Venezuela's state-owned oil concern. PDVSA described for the Conoco group the wonders of the Orinoco Belt, which held some 270 billion barrels of recoverable crude oil, a bonanza surpassing Saudi Arabia's 262 billion barrel reserves.

"Zemella told us that PDVSA was likely to receive approval to open this area to foreign investment," recalls Griffith. "He asked right then and there for concrete proposals to develop Orinoco, and he recommended we all go to lunch and begin discussions. The team looked at each other and decided lunch was a good option!"

Orinoco's crude oil was so dense and viscous it had defied attempts at commercial development for decades. "The stuff resembles asphalt," Griffith says. "Actually, one of its first uses was as a paving material in Chicago in the 1910s. PDVSA was looking for a way to produce this oil for about $3 a barrel, a very low price at the time for extra-heavy crude. We had no ready answers."

On the plane home, Griffith contemplated the challenge of economically developing the Orinoco Belt. "I said to Crude Oil Trader Barry Garvin, who was sitting next to me, 'I bet we could take that stuff and beat it up with our Lake Charles coker to make it economical,'" he recalls. "I sketched it out on a piece of paper, and when I got back to my office I pulled some numbers. It seemed like it would work, so I went around looking for someone who would agree it was a good idea."

After knocking on several doors, Griffith captivated both Colin Lee, executive vice president of downstream, and Jim Myers, general manager of petroleum coke and specialty products, with the idea. Within a few weeks, a new department — Downstream International Business Development — had been created. Several managers, including Gary Edwards, Mike Stinson, Rob McKee and Vice President of business development Don Unruh, lent their support to push the project forward.

Conoco was one of 22 oil companies at the time studying ways to develop Orinoco crude. Most of the companies examined expensive crude oil upgrading strategies, such as hydrocracking. Conoco had a less capital-intensive concept.

Griffith and the team presented Conoco's coking idea, which called for about a $1 billion investment by each partner. PDVSA was intrigued by the plan, especially since no other presentation had come in under $4 billion. The Venezuelan executives later met with Dino Nicandros, who was so impressed by their enthusiasm for the project he elevated it on his list of upstream priorities.

A letter of intent was signed in November 1991, and a joint feasibility study to develop Orinoco's extra-heavy oil reserves was completed the next July.

In late 1992, Conoco and PDVSA submitted a joint-venture proposal to Venezuela's Congress, which was approved in August 1993. Finally, on November 10, 1995, Conoco and PDVSA signed an association agreement. The project, wrote *Forbes* magazine, "dramatically shifts the balance of power in the oil industry." The business magazine noted that Venezuela's proximity to the United States "could stop all imports from the Middle East to this hemisphere."

The partners named their endeavor Petrozuata. The 35-year venture, in which Conoco has a 50 percent interest, took six years of intensive negotiations to structure. Venezuela's status as an emerging market, and the decades of exclusion of foreign involvement in the oil industry, had created nearly insurmountable odds for financing the project. Conoco's treasury department was given the difficult task of finding a solution.

"For a long time, it seemed that the project was not financeable, at least not in an acceptable way," says Miguel "Mike" Espinosa, then assistant treasurer of Conoco. "But the partners remained committed, and they continued, with the expectation that things would improve. And, in fact, they did."

Ultimately, the total cost of the $2.5 billion project was funded 40 percent by equity and 60 percent with debt. The debt consisted of a $1 billion bond offering and a $450 million syndicated bank loan facility. The financing was

Workers drilling in the Orinoco Belt in Venezuela drill and complete wells that produce some of the world's thickest crude oil. Conoco's proprietary technology is at work in every step of this integrated project, from producing to transporting to refining.

designed to be acceptable to lenders without the use of, or requirement for, political risk insurance, resulting in substantial savings in financing costs.

The bond issue was closed in June 1997, and the net result was the largest project financing in Latin America, with the lowest overall interest cost to date. A number of other project financing records were shattered, including the largest emerging-market project financing bond from a sub-investment-grade country. The bonds were rated higher than those of the Venezuelan government and were the highest-rated Latin American project finance bonds at the time. These precedents, and others, were cited by *Project Finance* magazine, which called the financing for Petrozuata "the deal of the decade."

With political and financial challenges surmounted, Conoco and PDVSA went after Orinoco's crude. Drilling would be a challenge, as would transporting and refining the oil. Conoco's expertise in well engineering and technology would be crucial to every phase of this fully integrated project. The company took advantage of the economic efficiencies offered by horizontal drilling technology. The higher per-well production rates achieved by horizontal drilling would minimize the number and cost of wells.

In August 1997, the first well was spudded in the Zuata field, the 55,000-acre drilling and production site. "It takes many more wells to produce extra-heavy crude oil than it does to produce lighter, more conventional crude that flows relatively easily through rock formations," says Jeff Stanat, technical services manager for Petrozuata. Over the life of the project, more than 500 horizontal wells will be drilled. To help push the slow-moving oil out of the ground, Petrozuata uses submersible pumps employing Conoco design features.

In August 1998, the first barrel of diluted extra-heavy crude was produced and began transport through a new 125-mile (201-kilometer) pipeline from Zuata to an industrial complex in the city of Jose on Venezuela's northern coast. It is the first private pipeline built in Venezuela in more than 20 years. Early production figures for the Petrozuata heavy oil project were 60,000 barrels per day.

At present, Petrozuata is in the process of completing a $1 billion upgrader at Jose. Conoco's delayed coking technology will be used to upgrade Orinoco's heavy crude into lighter synthetic crude and various by-products, including coke, sulfur, naphtha and liquefied petroleum gas. Roughly 60 percent of Petrozuata's output will be shipped from Jose by tanker to Lake Charles as feedstock for further refinement into high-grade fuels, lubricants and other products.

Petrozuata is vitally important to the future of both partners. Over the next 35 years, the joint venture is expected to produce roughly 2 billion barrels of extra-heavy crude — a production rate of about 120,000 barrels a day — helping to achieve Venezuela's ambitious goal for productivity and boosting Conoco's reserve base by 35 percent. A phrase coined by Mike Espinosa to describe the monumental project says it all: "Juntos saldremos adelante" — "Together we will come out ahead."

Looking back, Dave Griffith says it was the persistence of several individuals, including Bill Gormley, Tom Casbeer, Henry Van Wageningen, Paul Heard and Gary Poffenbarger, among others, who helped improve the odds and make the daunting project a success. "Unlike other oil companies, we've never been intimidated by the big projects, whether it's negotiating them, designing them or developing them," says Rob McKee. "We just keep chipping away."

To honor the success of Petrozuata, Griffith keeps a jar of Orinoco's unfinished crude in his office next to a jar of the finished syncrude. The contrast is startling...and satisfying.

Conoco's deftness at negotiating international deals, its masterful management of large projects and its estimable technology combined to make it a partner of choice for global exploration.

By the time rules against outside investment were relaxed as part of Venezuela's "Apertura" in 1995, Conoco was in the forefront of firms bidding for exploration rights. Of the ten blocks up for lease, two — the Gulf of Paria West and Guanare —

WORKING WITH JANE GOODALL

Jane Goodall began studying chimpanzees in Africa in 1960, revolutionizing perceptions of humanity's closest relative. Forty years later, Goodall, above, and her staff of researchers continue to contribute significant findings on chimpanzee behavior and social relations. Conoco partnered with the famed scientist to build a sanctuary for orphan chimpanzees in the Congo, a West African country in which the company had begun exploration. Conoco also showed what Goodall termed "real environmental responsibility" in this effort. "[Conoco's] teams walked through the forest," Goodall wrote in *Reason for Hope: A Spiritual Journey*. "There were no roads to mark their tracks. Far from the environmental watchdogs, in remote parts of Africa, Conoco nevertheless insisted on the same rigorous standards — with regard to personnel safety as well as the environment — that are enforced in the developed world." This testament, from such a respected advocate, ranks among the finest accolades ever given Conoco. Pictured below is Conoco Senior Vice President Mike Stinson with Goodall.

were granted to Conoco. In early 1999, Conoco made a discovery of potentially significant accumulations of oil in the Gulf of Paria West block. "The results are encouraging and reveal the potential for an important discovery, but it is still too early in the appraisal process to determine the real value," Rob McKee says. Conoco, operator of the project, has retained a 50 percent interest in the lease while farming in other partners.

Venezuela is an area with significant proven, discovered reserves. Its proximity to the United States presents many advantages, including the ability to leverage resources and realize both operational and cost efficiencies. Conoco already has a solid anchor with Petrozuata and a longstanding relationship with PDVSA as a trading partner. In Venezuela, Conoco can achieve integration of its operations — upstream, downstream, pipeline and marine. Consequently, Venezuela has great potential to become the company's third core area of operations, after North America and Western Europe.

In other areas spanning the globe, similar prospects burgeoned. The Atlantic Margin, west and north of the United Kingdom, offers hope as a future replacement for Conoco's North Sea oil and gas reserves. Conoco and three partners were awarded exploration licenses for two sought-after deepwater blocks in the area west of Great Britain's Shetland Islands, and Conoco was selected to serve as operator. Development, however, is challenged by an extremely inhospitable climate, including 100-mph winds and 100-foot seas. "A fairly sheltered bay has a constant 10-foot rise and fall," says Keith Webster in public affairs. "When some truly bad weather whips up, a 9,000-ton ship is tossed about like a leaf on a windy day — not that that deters us."

Elsewhere along the Atlantic Margin, Conoco holds significant acreage in the Vøring Basin in Norway and the North Rockall Trough and Porcupine Basin off the coast of Ireland. Conoco has amassed considerable geological and geophysical experience and understanding of this large petroleum province.

In Nigeria, Conoco's successful management of the Ukpokiti project again testified to its skilled project development and resourcefulness. In mid-1997, oil began flowing from the offshore Ukpokiti field. The *Independence*, a Conoco VLCC tanker that was due to be retired from service, was converted into a floating production, storage and offtake vessel (FPSO). This shaved millions of dollars off the cost of Ukpokiti's development when compared to that of a fixed platform. The *Independence*'s boilers were modified to operate on natural gas produced from the field, instead of conventional bunker fuel, another savings. The ship has 1.7 million barrels of storage capacity and also serves as an export terminal.

The field itself — where production is expected to top 20,000 barrels of oil per day at peak — was

The Ukpokiti field, off the coast of Nigeria, right, was developed using subsea wells and a Conoco tanker that was converted to a floating production, storage and offtake vessel (FPSO). The entire development offered a very low cost. Conoco's Nigerian staff, above, is small but highly professional.

The *Deepwater Pathfinder*, capable of drilling in waters as deep as 10,000 feet (3,048 meters), is the latest advancement to come from Conoco in the area of deepwater exploration and production. In June 1999, *Fortune* magazine called the double-hulled drillship "Big Oil's Big Boat," noting its unique hybrid status as "half-ship/half-rig." The sophisticated drilling riser is deployed through the open "moon pool," right. Not only does the $265 million vessel incorporate the highest standards of safety, reliability and environmental protection, she also offers unique flexibility. The *Pathfinder* can simultaneously drill and test, and offers storage for up to 100,000 barrels of oil. The deck carries enough pipe, casing and riser, opposite, to drill two wells.

developed using low-cost, fit-for-purpose subsea wells and has very low operating costs. "While many of our competitors have gotten out of the tanker business, Conoco's marine department is capable not just of low-cost, environmentally safe shipping, but it is also able to see and create value for upstream projects," Edwards says.

Following the success of upstream's Polar Lights project in Russia, Conoco signed an agreement in March 1998 with Lukoil — the largest Russian oil company — to study development of reserves in the 1.2-million-acre region of Russia known as the Northern Territories. The area is believed to contain more than 1 billion barrels of oil and 2 trillion cubic feet of natural gas. When Polar Lights began production in 1994, Archie Dunham said, "Russia could fundamentally change the size of the company." As the new deal with Lukoil takes shape, Conoco is well-positioned for the future in Russia.

Other promising partnerships were forged in new energy frontiers, such as the country of Trinidad and Tobago. In 1997, Conoco signed a production-sharing contract for two large blocks off Trinidad's east coast, and drilling began in 1999 in both areas.

In North America, Conoco has focused much of its exploration investment in the deepwater Gulf of Mexico, estimated to hold more than 10 billion barrels of oil. The Gulf was once a hot spot for oil production, but development stagnated as companies

reached the limit of how deep they could drill with existing technology to produce the oil economically. Indeed, by the mid-1990s, the Gulf became known as "The Dead Sea."

With its significant expertise in deepwater exploration, drilling and production, Conoco has helped revitalize the Gulf of Mexico. The company today is the seventh-largest leaseholder in the Gulf, with 295 deepwater leases, some of which lie in more than 10,000 feet (3,048 meters) of water. Among these is the Ursa field, with a peak production expected to reach 150,000 barrels of oil and 400 million cubic feet of gas a day by 2001. Conoco holds a 16 percent interest in Ursa. "Our current portfolio of producing properties in the Gulf is slanted about 80 percent toward gas and 20 percent toward oil," says former Gulf Region Manager Bill Brister. "So Ursa production represents a major addition to our oil production — 75 percent, to be precise."

Although Conoco has a relatively small interest in Ursa and was not the operator of the project, its technological expertise was instrumental in building the deepest tension-leg platform in the world today. (Conoco had installed the Gulf's first TLWP, or tension-leg well platform — Jolliet — in 1989.) The 97,000-ton Ursa TLP is connected to the ocean's floor, 3,800 feet (1,158 meters) below, by 16 tendons, each three-quarters of a mile long and

INDEPENDENCE 251

The *Pathfinder* drillship, left, is a ship with a grand scale: 721 feet (220 meters) long, 144 feet (44 meters) wide, 66 feet (20 meters) deep and with an overall displacement of 103,000 deadweight tons. The vessel was christened in Korea in late September 1998. Pictured at the ceremony (from the left) are Paul and Penny Lloyd, Rob and Ann McKee and Mr. and Mrs. J. W. Kim, representing R&B Falcon, Conoco and Samsung Heavy Industries, respectively. Eight months after beginning exploration in February 1999, this vessel had an amazing track record — two discoveries in the Gulf of Mexico on its first two exploratory wells.

INDEPENDENCE 253

ALLEN D. GAULT

Allen Gault loves nothing more than jumping out of a plane with a parachute on his back. "I guess you could say I'm a guy who knows where he wants to get and has the skills to get there, but enjoys most the journey along the way," says Gault, a senior staff engineer who came to Conoco in 1991. Gault specializes in technology development and transfer, most recently as the project manager for the SubSea MudLift Drilling Project — to enable drilling in 10,000 feet of water. Gault recruited 20-odd companies, put together a feasibility study and raised $1 million for the first phase of work. He says the technology would have continued to collect dust were it not for Conoco. "Most companies just focus on the technology development, not its transfer. As a result, they sometimes fail. At Conoco, there is an understanding that implementation is just as important as invention. That turns me on."

weighing 1,000 tons. With a total deck height of 485 feet (148 meters) — more than 48 stories high — Ursa is the largest structure in the Gulf, visible on a clear day from more than 100 miles away.

"We cut our teeth in the Gulf in the early 1950s and were, in fact, one of the first companies to drill out of sight of land," says McKee. "We consider the Gulf the foundation of our offshore marine and deepwater expertise. Moreover, it is a prolific basin that was underexplored. Given our history there, our refinery in Lake Charles on the shores of the Gulf and our deepwater drilling strengths, we were a natural to participate in the reopening of this area."

Success in the Gulf's extreme water depths, though, would require enhanced deepwater drilling capability. In 1996, Conoco announced plans for a joint venture with R&B Falcon, a major offshore drilling company, to construct a giant, double-hulled drillship — the *Deepwater Pathfinder*. The ship, which can drill in waters as deep as 10,000 feet, accommodates cost-effective ultra-deepwater exploration and production in the Gulf.

Korea's Samsung Heavy Industries, who built all of Conoco's double-hulled tankers, constructed the *Pathfinder*. The vessel is an astonishing sight — 721 feet long, 144 feet wide, 66 feet deep (220 meters long, 44 meters wide, 20 meters deep) and with a displacement of 103,000 tons. Not only does the vessel incorporate the highest standards of safety, reliability and drilling efficiency, she also offers unique flexibility, can simultaneously drill and test, and can provide storage for up to 100,000 barrels of oil.

Pathfinder uses a satellite-guided system of thrusters to hold it precisely in position when drilling, despite hurricane-force winds, waves or currents. "The drillship enables us to more economically reach oil and gas reserves lying beneath the deepest seas, and to do so more safely than ever before," says Antonio Valdes.

The drillship was christened in Korea in late September 1998 and then sailed toward the Gulf of Mexico, where it drilled its first exploratory well in early 1999. Over the next five years, it will work for Conoco exploring the company's deepwater tracts. A second, identical drillship, the *Deepwater Frontier*, was completed in early 1999. The *Frontier* will operate in deepwater tracts in the Atlantic Margin, Norway, Nigeria and New Zealand, as well as the Gulf of Mexico. "We didn't invest in the drillships to be in the drilling business, but to enable the orderly exploration of our holdings in the deepest waters of the world, which in turn grows this company," says McKee.

"We drew on Conoco's double-hull tanker technology, on Samsung's shipbuilding expertise and on our deepwater drilling expertise to come up with a winning design jointly with R&B. What we have now are the most sophisticated drillships on the planet. I sit here in my office and long to be aboard

The newest addition to Conoco's branded retail strategy in the United States is "breakplace." Redesigned convenience stores featuring updated decor, fresh-ground coffee in a host of flavors and a piping hot bakery selection achieve a restaurant feel.

them," adds McKee, whose career began in the Gulf of Mexico as an offshore drilling foreman.

At the news conference announcing the decision to build the first drillship, Archie Dunham said, "Having a drillship solely dedicated to our deepwater portfolio gives us the tools and expertise to fully explore our deepwater acreage around the world."

Throughout the 1990s, Conoco turned potentially profitable ideas into valuable projects. Dunham was determined to see his dream of doubling the company's value to $30 billion become a reality. Perhaps the most remarkable aspect of all these endeavors is the company's successful management of so many large, complex projects at once. Of the six largest projects Conoco has undertaken in the past several years — the deepwater drillships, the Melaka refinery, the Petrozuata project, Britannia, the Lake Charles lube oil hydrocracker project and the Lobo acquisition — any one would testify to the company's unique skills in project management. But to have undertaken all of these and more in a decade, and for all to succeed, sets Conoco above the industry in a class by itself.

It was an exciting time in the company's history. Yet, the most startling and exciting event of the decade was saved until the final hours of the twentieth century.

Since 1981 — when Conoco's name was erased from the New York Stock Exchange following its merger with DuPont — Conocoans around the globe shared a dream that their company would one day again stand on its own. "I decided that within the first three years of my presidency, I would make a move to separate Conoco from DuPont," says Archie Dunham. In 1998, a confluence of events convinced him the time was ripe.

"The original justification for the merger had evaporated," Dunham says. "The energy crisis had disappeared, and DuPont no longer needed Conoco's earnings to 'cut the trough off their earnings curve.'" Moreover, DuPont had stated its desire to transform itself into a life sciences company, one focused more on biotechnology than on specialty chemicals. By divesting Conoco, DuPont could reap substantial capital to invest in a makeover. "As a member of DuPont's board, I saw that Conoco did not fit in with the strategic refocus of DuPont," Dunham says. "A spin-off appeared appropriate and inevitable."

Three years earlier, Dunham says, Conoco would not have been ready. Since then, the company had undergone a "fundamental change" to focus its diverse portfolio and drastically reduce costs. In addition, Conoco's upstream projects in Venezuela, the North Sea, Malaysia and Russia, its downstream ventures in Eastern Europe and Southeast Asia, and a booming lubricants business had positioned it for solid future growth. "In the first quarter of 1996, I asked Tom Henkel and Jim Allison to seriously evaluate the feasibility of separating Conoco from DuPont," Dunham recalls. "They prepared a highly confidential memorandum for my review which suggested there was a lot to be gained."

"The idea had been floating around below the surface for some time," Allison, manager, financial analysis, says, "but the question always was, is this the right time for both Conoco and DuPont? The answer required analysis of the opportunities and risks for both companies. We determined that there were significant new international opportunities for Conoco. Meanwhile, DuPont was moving toward expanding into life sciences. Both opportunities required substantial investment, making it unlikely that they could be pursued simultaneously. Allowing Conoco and DuPont to separate and pursue their respective goals, therefore, was a way for shareholders to retain access to the potential value creation from both paths."

In 1996, Dunham quietly went outside the company for additional advice, engaging analysts Mike Mayer, Bob Israel and Mike Wagstaff of Schroders to counsel him on the feasibility of separating Conoco from DuPont. In 1997, Dunham asked the team to update their report. Mayer was convinced the stock market would respond enthusiastically to a Conoco initial public offering (IPO). "In upstream, Conoco had radically restructured its operations and substantially improved profitability," Mayer says.

"Looking forward, its oil and gas production was expected to grow by almost 33 percent from 1997 to 2000. This was the best rate among the majors and about 50 percent better than the group average. In downstream, its operating costs had been reduced by about one-third, while its refining throughput was up 20 percent." It was clear that Conoco's competitive position was very strong.

Buoyed by the Schroders report, Dunham sought additional outside analysis from McKinsey & Company and Cambridge Energy Research Associates. "They pointed out that a window of opportunity in the oil industry existed in the ensuing three years to make significant foreign-asset investments," Dunham says. "Twenty-six countries had opened their oil industries to foreign investment since 1993, and Conoco would need to make huge investments in selected new countries before the window closed." The timing for a separation had never been better. Now all Conoco had to do was convince DuPont's executive management and its board of directors.

Dunham first broached the subject of a separation to DuPont's future CEO, Chad Holliday, at the Purple Sage Ranch in Texas in September 1997. "I remember walking down the airstrip with Chad at eleven o'clock at night telling him what I wanted to do and why," Dunham says. "We talked about DuPont's need to shift its portfolio toward life sciences and biotechnology. No

It's considered one of the world's most advanced and efficient power plants, a $200 million, natural gas-fired unit that can produce approximately 440 megawatts of electricity and up to 1.1 million pounds of steam per hour. Ingleside Cogeneration, a joint venture of Conoco Global Power and Occidental Petroleum Corporation, was completed in 1999.

INDEPENDENCE 257

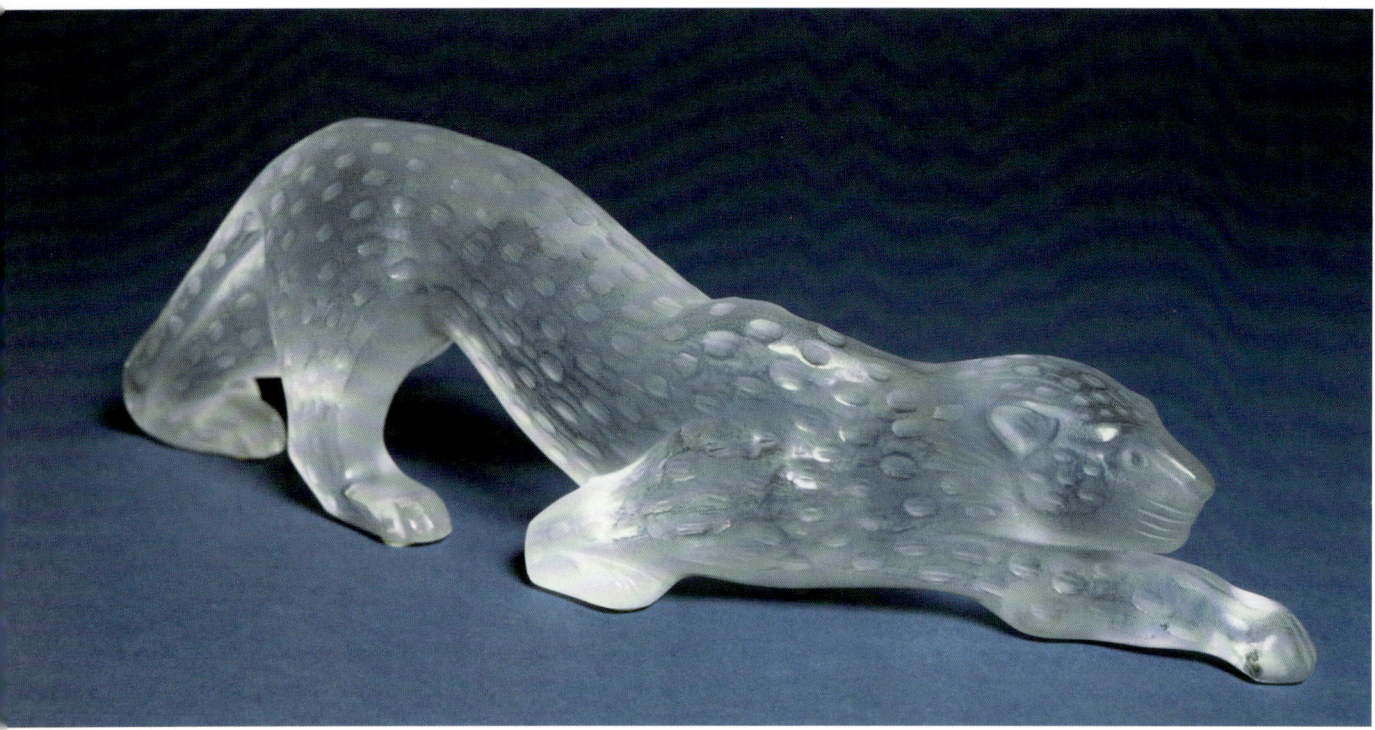

The crystal statuette above commemorates the company's code name for separating Conoco from DuPont — Project Leopard. Archie Dunham selected the name because it conveyed characteristics that were in sync with the company, such as "fast and nimble," he says. "We are a company that can change direction quickly, like a leopard. We're decisive!" When DuPont approved the deal, Dunham sent a confidential e-mail to management confirming, "The leopard is at the door."

decisions were made then, but I was elated that, philosophically, we were in agreement on what had to happen next."

In February 1998, Dunham presented his idea to DuPont's office of the chief executive, explaining the rationale for a Conoco spin-off. He outlined Conoco's global vision and growth opportunities in South America, Southeast Asia, Russia, the Middle East, North America and Europe. He demonstrated how the company's growth plan would add substantially to future earnings. In summary, he emphasized that separating Conoco from DuPont would be in the best interests of both companies and would substantially benefit shareholders.

On April 9, 1998, Dunham made a second presentation to DuPont's office of the chief executive. He recommended they "create two great companies, not one strong DuPont and a weakened Conoco." Says Dunham, "They agreed."

On April 29, Dunham and Holliday made a joint presentation to DuPont's board of directors recommending an IPO of Conoco. On May 10, the board agreed a reborn Conoco would benefit both companies and their shareholders. After 17 years as a DuPont subsidiary, Conoco was moving toward becoming an independent oil company once again.

The news was released to the world the following morning. "I wish I could open my jacket and show off a T-shirt that says 'Independence Day!'" a visibly elated Dunham exclaimed at a news conference, referring to the hit Hollywood movie with the same title. *Conoco World* printed a special edition, with big block letters proclaiming in a headline: "An exciting, historic day for Conoco." The words were the CEO's.

DuPont planned to offer 20 percent of Conoco's common stock to the public in an IPO, making it one of the largest such deals in history. The company also announced its intention to divest the remainder of Conoco within the next 12 months. The financial press responded positively. "This is going to work. This is big. [Conoco] is a real company," Vincent Slavin, an analyst tracking IPOs for Cantor Fitzgerald Inc., commented in the *Wall Street Journal*.

Although ecstatic at the board's decision,

employees at every level of the company knew the coming months would be critical to the success of the IPO — and they all pitched in to make it work. The legal and finance groups carried much of the direct load. Conoco CFO Bob Goldman was instrumental in assembling the financial transaction. "We worked with Conoco lawyers and legal assistants in the U.S. and Europe, finance analysts from operations and business groups, and literally scores of support people virtually nonstop from May till October," says Goldman. "We spent thousands of extra hours and gave up holidays and vacations to complete the due diligence work, determine a corporate structure for the new Conoco and prepare a registration statement for the Securities and Exchange Commission." Employees at every level indirectly contributed by keeping the company going in the midst of many distractions.

Dunham would later recall this period as the most stressful of his life. "The media printed numerous rumors, one of which was that Conoco was about to be purchased by Elf Aquitaine of France," he says. "Unfortunately, we were in a Securities and Exchange Commission (SEC)-mandated 'quiet period' because of the IPO, and I could not comment on the rumors, which were completely false. It was exasperating. I felt the pain of employees worried about the company and yet could do very little about it."

In September, Conoco filed with the SEC to issue 150 million shares in an IPO. But obstacles

THREE HISTORY-MAKING
FINANCIAL DEALS

Conoco was reintroduced to the worldwide financial community with a three-part financing program, including a $4 billion global bond issue. When issued, it was the largest-ever bond offering by a U.S. energy company. This was followed by a borrowing arrangement with 33 domestic and international banks that would be utilized to support a $2 billion commercial paper program. Thus, with borrowing facilities totaling $6 billion, Conoco paid off its remaining indebtedness to DuPont, and set the stage for the final phase in the separation of the two companies. Each of the three phases — IPO, financing and split-off — involved Conoco in a transaction that was the largest of its kind in the history of U.S. finance. Pictured is a replica of the new Conoco shares that were issued.

INDEPENDENCE 259

beyond Conoco's control lay in the way. Says Dunham, "The stock market began moving like a roller coaster, and oil markets took a nose dive. In short, a horrible business climate threatened the entire plan." Six other companies had canceled IPOs in the weeks before Conoco's announcement, and even the *Wall Street Journal* questioned the wisdom of the company's timing. "Some analysts remain highly skeptical of the success of a $3 billion to $4 billion initial public offering — one of the largest ever — at a time when world financial markets are in turmoil and the appetite for IPOs has all but dried up," the *Journal* reported.

Dunham, however, saw opportunity amid the tumult. He assembled his management teams and set out on a "roadshow" — a full-out effort to sell scores of institutional investors and stock analysts on Conoco. The trek would take two groups — Dunham, Edwards, John Kemp and Tom Henkel; and McKee, Bob Goldman and Jim Nokes — to 47 cities in 10 countries in three weeks. This contingent made 120 speeches during that span, sometimes seven or eight in a single day.

Conoco's story stirred Wall Street's giants. The results of the trip were phenomenal. Despite low oil prices and a volatile stock market, investors responded with enthusiasm to the offer. "We took the Conoco story public and told it so well that we oversold the issue by a multiple of four," Dunham notes with pride. In fact, 70 percent of the investors who had heard Conoco's story purchased the stock. It was an astonishing accomplishment given that on average, investors purchase stock of only 20 to 30 percent of the IPO opportunities presented to them. With the IPO substantially oversold, DuPont upped the ante. Instead of a 20 percent stake, the IPO would take 30 percent of the company public.

Conoco stock began trading again on October 22, 1998, priced at $23 a share. It was the only U.S. IPO to be priced that month, yet one look at the stately facade of the stock exchange as the opening bell rang left no doubt that the company was back. A huge red and white banner bearing the Conoco capsule was draped from the building's Corinthian pillars. On the street below were old-fashioned gas station replicas adorned with the words, "We're back!" and "We're pumped!" — a message that had been hammered home by the events of the past several months. The *Journal*'s assumptions proved wrong. The IPO was greatly oversubscribed. At a whopping $4.4 billion, it was the largest in U.S. history at the time.

That evening, some 100 employees and invited guests in formal attire attended a reception on the trading floor of the exchange, where Conoco's stock had made its triumphant return using a new symbol, "COC," honoring the name it had held for so many years — the Continental Oil Company. At the dinner, Conoco's management team presented Dunham with a treasured gift, an authentic U.S. Marine Corps

A DAY IN NEW YORK CITY

The Big Apple was the site of several stirring celebrations to honor the return of Conoco to the New York Stock Exchange and its new life as an independent company. Among the Conocoans celebrating in New York City were 15 employees who were awarded the trip as part of a companywide essay contest in which they were asked to describe their personal feelings about Conoco independence. From Russia, Venezuela, Dubai and elsewhere, they responded in an emotional outpouring of words. Titie Wibipriatno from Indonesia commented, "I am very proud to work for a company that...is well-respected by my government for its core values." And veteran Conocoan Mike Rooney in Venezuela recalled an early spring blizzard in March 1975, in Cheyenne, Wyoming, where he was stationed upon joining the company. "My wife and I were so excited about our future with Conoco that running out of fuel in a blizzard did nothing to dampen our enthusiasm for the adventure! We wait for this new adventure as if we were breathless, excited children on Christmas morning!" Contest winners are featured in the photos to the right, visiting the Statue of Liberty and other famed New York attractions.

INDEPENDENCE 261

A gala reception was organized on the trading floor of the New York Stock Exchange to commemorate the IPO. Some 100 Conocoans and invited guests — some in native regalia, left — attended the dinner, at which Conoco's CEO Archie Dunham was presented an authentic U.S. Marine Corps sword. The sword was inscribed: "To Archie W. Dunham, for leading from the front in Conoco's battle for independence."

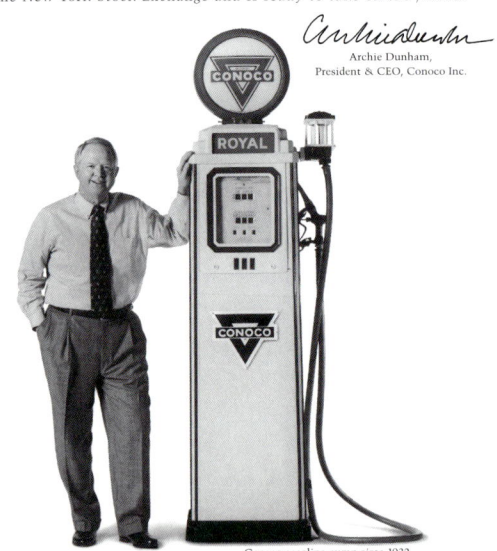

sword, just like the one he had been authorized to wear when he was commissioned in 1960. On the sword were the words: "To Archie W. Dunham, for leading from the front in Conoco's battle for independence."

Amid the Conocoans celebrating at the exchange that day were 15 employees who were winners of a special company essay contest, in which they were asked to describe their personal feelings about Conoco independence. From Russia, Venezuela, Dubai and elsewhere, they responded in an emotional outpouring of words. "Nine years ago, my country was becoming independent," wrote Lucie Melicharova, a human resources advisor from the Czech Republic. "I fought for its independence and then celebrated the so-called 'Velvet Revolution' in one of the greatest moments of my life. Today, I strive for you to be independent. I look ahead with great expectations as I know what it takes to become independent. It takes a lot of courage and strong will. You must have great people who trust you and share the right values. You, Conoco, have it all."

"The success of the offering spoke volumes about the financial community's confidence in Conoco's ability to grow rapidly and generate superior returns on its shareholders' investments," Dunham says. "From my perspective, it also was a vote of confidence in our 15,000 dedicated employees worldwide."

On Friday, October 23, 1998, Dunham — in a red cap sporting the Conoco logo and stock symbol

On October 23, 1998, Archie Dunham partook of a New York Stock Exchange tradition. He and his wife, Linda, were invited to ring the opening bell at the New York Stock Exchange to celebrate the triumphant return of Conoco as an independent, public company. Joining the couple is Richard Grasso, chairman and CEO of the NYSE.

Thousands of company employees all over the world watched the ringing of the bell at the New York Stock Exchange signaling the birth of a newly independent Conoco. The employees gathered in celebrations from Hamburg to Houston to share some victory cake or otherwise let loose their emotions. The event was broadcast via live satellite from the floor of the exchange and hosted by Conoco's communications staff. On August 9, 1999, DuPont's ownership of Conoco came to an official end. Conoco faced the future on its own.

and with his wife, Linda, by his side — rang the opening bell at the New York Stock Exchange, signaling the start of the day's trading and the start of a new era in Conoco's history. Thousands of company employees from Houston to Hamburg watched the event via live satellite broadcast from the floor of the exchange hosted by Conoco's communications staff. They rejoiced over the news of an independent Conoco. Among the hundreds of messages received by Dunham in the last weeks of October 1998 was this one from a retiree: "I must admit I do cry during 'The Star Spangled Banner' at football games. And even though I'm no longer a Conoco employee, those same tears of pride welled up when I saw COC back up on the New York Stock Exchange."

Another wrote, "This is the first time I've sent you a note, and in truth we'll probably never meet. (But) you have no idea how many employees — indeed families — are relying on your ability to guide Conoco to the successes we deserve. Thank you for carrying this heavy burden."

On August 9, 1999, DuPont's ownership of Conoco came to an official end. DuPont stockholders by the thousands took the company up on its offer, receiving 2.95 shares of Conoco Class B stock for each share of DuPont stock they tendered. The offer, for up to 148 million shares of DuPont stock, was oversubscribed by nearly 2.5 times, with just over 353 million shares tendered in all. In a special issue of *Conoco World*, bearing the headline, "On our own and raring to go!" Dunham welcomed employees to a new, independent Conoco, "a company that faces the future as an aggressive, confident competitor in the global energy industry. It's going to be an exciting ride. By keeping a steady hand on the wheel and doing what we do best, we will be first to cross the finish line…to get to the future first."

Conoco, now passing the 125-year mark, is reinforcing its position as a truly great company that, by all accounts, is indeed getting to the future first. The small marketing enterprise that sold kerosene, candles and wax during the Industrial Revolution and built its promise on an innovative oil barrel attached to a rocker, is now an international integrated oil company with almost three billion barrels of oil and gas reserves and a broad line of high-quality products. The horse-drawn tank wagons have given way to gargantuan double-hulled tankers, drillships and enough pipeline to circle the Earth. The English cottage gas stations have been replaced by today's modern outlets — some complete with unmanned functions, others with comfortable convenience stores and enough gas to power millions of cars. The spirit of the Continental Soldier resides still in the Conoco capsule.

Today's Conoco is a respected S&P 500 index-traded company ranked tops in oil exploration and production. Those early oil hunters, Marland and McCollum, would be proud to learn that in 1998, for the sixth consecutive year, Conoco replaced more reserves than it produced. At year-end, its reserves were at 2.6 billion barrels of oil equivalent — the highest level in 20 years. Evidently, their Midas touch has been passed on to a new generation of explorers, men and women skilled in the art of developing energy.

As its employees bask in the promise of a new Conoco, the once-modest company founded in Ogden, Utah, in 1875 seems assured of a prosperous future. Its core values of operating safely, protecting the environment, adhering to high ethical standards and valuing all people have proved to be a solid foundation upon which to build a new legacy. "These values are more than just good intentions or slogans," Archie Dunham says. "We're committed to them not only because they're morally right, but also because our reputation is critical as we seek to build new relationships around the world."

And so our story that began with two companies, one in the Rocky Mountains and the other in the Sooner lands of the Southwest, continues with one company, strong and independent, known the world over. Its singular record of achievement is a testament to the insight, ingenuity and exertions of the many thousands of employees who, for the past century and a quarter, have called themselves Conocoans.

Timeline

1875
On November 25, Isaac Elder Blake, along with four partners, launches the Continental Oil and Transportation Company in Ogden, Utah, to distribute petroleum products to the West.

1876
Continental opens an agency in Denver to market oil, forming the Continental Oil and Transportation Company of Colorado.

1877
Continental establishes sales agencies in Utah and Montana. The Continental Oil and Transportation Company of California is incorporated.

1878
Continental's top products are listed as oil for lamps, candles and waxes. The company begins limited overseas marketing in Canada, Mexico, the Hawaiian Islands, Samoa and Japan.

1883
Continental builds the first pipeline in California, from outside Santa Clarita to Ventura.

1884
Continental moves its headquarters from Ogden, Utah, to the McPhee Building, later renamed the Continental Oil Building, in Denver. The Continental Oil Company of Iowa is incorporated.

1885
Continental's affiliation with Standard Oil officially begins January 1. Early advertising strategies, portable oil barrels and double-holed barrels are developed.

1890
Continental's major markets include Colorado, Wyoming, Montana, Utah and New Mexico.

1893
Henry Morgan Tilford becomes Continental's president.

1906
Continental lists 60 distribution points throughout the United States and reports to have 98.9 percent of the Western U.S. kerosene market. The company is still predominantly a marketer of oil, although records indicate a burgeoning interest in field operations.

1907
Edward T. Wilson becomes president and CEO of Continental.

1908
Ernest Witworth Marland secures his first oil lease from the 101 Ranch in Ponca City, forming the 101 Ranch Oil Company.

1909
Continental develops the gravity feed dispensing system for gasoline, an increasingly important product. The setup is considered the first true filling station in the West.

1910
There are 460,000 cars in the U.S., up from 300 in 1895. Continental invests in motor trucks to transport fuel to retail outlets.

1911
Marland strikes first oil June 27 on Ponca land — the Willie Cry well.

The U.S. Supreme Court orders the dissolution of the Standard Oil Trust.

1913
Continental Oil Company of Colorado is incorporated once again as an independent company on March 31. There are 1.7 million cars in the U.S., a ready market for the company's "Conoco" brand gasoline.

1916
Continental enters crude oil refining and production, purchasing the United Oil Company and its Florence, Colorado, refinery.

The 101 Ranch Oil Company is producing oil from three Oklahoma wells, primarily for sale to refining companies. The Elk Basin Petroleum Company is established and begins a string of acquisitions that would eventually lead to Continental.

1917
The 101 Ranch Oil Company is reorganized as the Marland Refining Company.

Continental begins an agreement with Yellowstone National Park to be the exclusive supplier of gasoline to the park, an agreement that remains in effect through the end of the century.

1918
Marland Refining completes a new refinery and tank farm in Ponca City, Oklahoma, and enters the booming retail market in Oklahoma, Kansas, Arkansas, Missouri, Nebraska and Iowa.

1919
The red triangle is adopted as the symbol of Marland Refining and, later, the Marland Oil Company.

1920
Marland's aggressive exploration strategy pays off with the discovery of the huge Burbank field in California on May 14.

1921
Marland Refining pioneers seismographic exploration in the U.S., discovering the Tonkawa field in Oklahoma and leasing acreage in Mexico — the beginning of the company's international exploration effort.

1922
The Marland Oil Company is formed as a holding company for Marland's various subsidiaries.

1923
E. W. Marland and J. P. Morgan meet for the first time to discuss financing for the rapidly expanding Marland Oil. The company acquires the rights to the germ process, a method for improving lubricating oils.

1924
Elk Basin Petroleum, renamed Mutual Oil, initiates a merger with Continental, creating a new company managed by Mutual Oil executives but bearing the Continental name and a new president and CEO from its ranks — C. E. Strong.

1925
Marland Oil discovers the Howard-Glasscock and Big Lake fields in Texas.

1926
Continental purchases a refinery in Wichita Falls, Texas. Marland acquires the Sealand Petroleum Company to market petroleum products in the United Kingdom.

Marland Oil has 5,000 tankcars and 600 service stations to transport and market its products in all U.S. states and in 17 foreign countries. Marland Oil strikes a deal with Hudson's Bay Company to explore for oil in Canada.

1988
Conoco develops the "V" fields in the North Sea and the Lincolnshire Offshore Gas Gathering System (LOGGS) to transport gas from the "V" fields to the Viking terminal at Theddlethorpe. The project is christened by Prime Minister Margaret Thatcher. Conoco (U.K.) Limited is the only offshore petroleum operator to receive the British Safety Council's Sword of Honor, given to the 30 safest companies in the world. Conoco inaugurates the annual Conoco Rodeo Run, which provides support for the Houston Texas Livestock Show and Rodeo Scholarship fund.

1989
Conoco completes the first ever tension-leg well platform (TLWP) at Jolliet field in the Gulf of Mexico. Conoco and partners form Phoenix Park Gas Processors Limited to construct a gas plant in Point Lisas, Trinidad.

1990
Conoco Indonesia's Alu Alu well hits gas in the West Natuna Sea. The new field is named Belida. Dino Nicandros announces nine new environmental initiatives, including the groundbreaking decision to use only double-hulled tankers to transport petroleum.

1991
Conoco Norway Inc. received permission from Norway's parliament to develop the giant Heidrun oil and natural gas field in the Norwegian Sea. Conoco opens Jet outlets in East Germany.

1992
Conoco Project 2000, an ambitious upstream reengineering initiative is announced. The company reevaluates its portfolio, shedding several assets. Additional restructuring programs are also undertaken, including Global Excellence for international upstream the following year and Target 2000 for downstream. Conoco and Chevron reach an agreement to operate jointly the Britannia field in the North Sea. Conoco reaches the first ever agreement with a Russian state oil company to develop a new field — the Ardalin field in northern Russia. The joint venture is called Polar Lights.

1993
Conoco begins to expand its retail presence in Eastern Europe, opening retail outlets in the Czech Republic, Poland, Hungary, Turkey and Solvakia. Conoco opens its first Jet station in Thailand and begins to develop downstream activities in nearby markets.

1994
Conoco signs an agreement with a Malaysian partner to build a refinery near the Malaysian town of Melaka. Conoco reaches a historic joint venture agreement to develop extra-heavy oil in the Orinoco belt in Venezuela, the first such deal in Venezuela in 20 years. Excel Paralubes, a $500 million joint venture with Pennzoil to produce high-quality base oils, is announced.

1995
Following U.S. government sanctions against Iran, Conoco withdraws from pioneering plans for development of the Sirri project in that country. Archie W. Dunham is named president and CEO. Conoco and two partners purchase a 49 percent interest in two Czech refineries in Kralupy and Litvinov. The giant Heidrun platform in the North Sea is christened. The Conoco Challenge employee bonus program starts.

1996
The Ursa platform in the Gulf of Mexico, in which Conoco holds 16 percent interest is completed as the world's deepest TLP. A new corporate structure is announced, starting a "fundamental change" at Conoco. A joint venture with R&B Falcon to construct double-hulled drillships is announced.

1997
Conoco acquires the Lobo natural gas field in South Texas, making it the second-largest natural gas producer in the state. The first well is spudded in Venezuela for the Petrozuata project. A new methane-processing plant fed from the Heidrun field opens in Norway.

1998
Conoco announces a deal with Lukoil in Russia to develop 1.2 million acres in Northern Russia for oil production. The Melaka refinery comes on-stream. Petrozuata hits first oil. Production begins from the Britannia field. For the sixth consecutive year, Conoco replaces more reserves than it produces. Conoco tops American Petroleum Institute's safety ranking again, more frequently than any other oil company during the 1990s. Recognizing Conoco's strong industry position and increasing opportunities for investment, DuPont announces an initial public offering of 20 percent of Conoco on May 25. After a whirlwind road show to win the support of investors, Conoco's IPO is the largest to date. Conoco stock begins trading again on October 22, 1998, priced at $23 a share.

1999
Conoco's deepwater drillship drills its first well. The number of Jet stations in Thailand tops 100. The simultaneous success of several major projects gives Conoco top ranking in oil exploration and production by Standard and Poor's. On August 9, Conoco officially becomes an independent company again.

2000
Conoco begins its 125th year in business.

ABOUT THE AUTHOR

Russ Banham is a Pulitzer-nominated journalist, author of five books and veteran writer of more than 2,000 articles published in a wide variety of business publications, including *Forbes*, *CFO*, *Financial Times* and *Venture*. Banham's book, *Coors: A Rocky Mountain Legend*, a biography of the Coors brewing dynasty, reached number four on the Denver Post's nonfiction bestseller list. His adaptation of four short stories by Anton Chekhov into the play, *Romance with Double Bass* (which his wife, Jennifer Sue Johnson, co-adapted) was produced by Seattle's Book-It Repertory Co. in 1999. Other books by Russ Banham include histories of USF&G and Boatmen's Bancshares and *A Primer on Insurance*. Banham also has written several video documentaries. He has two graduate degrees in the fine arts, both from the University of Montana, where he taught for three years as a Javits Fellow. Banham and his wife are currently at work on a theatrical adaption of Edith Wharton's novella, *Ethan Frome*. A father of two children, Isabelle and Charlie, Banham works from homes in Missoula, Montana, and Snoqualmie, Washington.

Index

Bold listings indicate illustrations.

A

Advance Exploration Group, 119, 145, 147
Africa, 137, **137**. *See also specific countries*
Agip, 217
Agroforestal, 244, **244**
Alaskan North Slope, 164 **164-165**, 165
Albequerque, New Mexico, marketing department, 83
Alfol, 120
Allison, Jim, 256
Alu-Alu E-1 well, 189
Amerada Petroleum Corp., 108, 109
America, 148
American Agricultural Chemical Company, 121
American Petroleum Institute, 180, 189, 234
Amoco, 132
Anatrol process, 121
Angola, 171
ARA, 145
Ardalin field, **199**, 200, **200-201**, 201, 202
Argentina, 113
Arrow, 116
Arkhangelskgeologia, 201
Artesia, New Mexico, refinery, 114
asphalt, 42, 213
Atlantic Margin, 248, 249, 254
Atlantic Refining, 104
Austria, 115, 145
automation. *See* computers
automobiles, 29, 36, 37, 40, 41, 47, 49, 51, 53, 75, 76

B

Bailey, Ralph E., 141, 142, 145, 150, 151, 154, 155, 158, **158**, 159, 160, 163, 178, **178**
Baker, Bryant, 62, **62**, 65
Baltimore Canyon Basin, 147
Baltimore, Maryland, refinery, 73, 76, 103, 121
Bangalore, India, development office, 220
Barnes, John, 166
Batson, Ed, 107
Belgium, 114, 145, 172, 173
Belida field, 189, 234
Bell, Charles, 108
benzine, 19, 36, 42. *See also* fuel: gasoline
Big Lake field, 61
Billings, Montana, refinery, 103, 114, 147, 148
Blackshear, Jack, 216
Blake, Isaac Elder, 14, 15, **15**, 17, 19, 21, 22, 31, 32, 33
Blauvelt, Howard W., 138, **138**, 141, 147, 150, 196

Bois d'Arc Creek, 28, 29
Bond, David, 216
Boston-Wyoming Oil Company, 53
Bowden, Connie, 193
Bowden, Kent, 193
Bowler, Dave, 145
Boulton field, 231, 233
Bradshaw, Terry, 175, **175**, 178
Braggs, Avis, 234, **234**
Braly, Bernie, 108
Bratton, Caroline, 213
breakplace, 255, **255**
Brent pipeline system, 145
Brister, Bill, 251
Britannia, 131
Britannia field, 225, 228, 232, 234
   development, 226, 231
   drilling teams, 228, **228-229**
   negotiations, 226
   platform, **226**, 227, **227**
British Gas, 166
British National Oil Corporation, 145
British Petroleum, 164, 166
Britton, Mike, 202, **202**
Bronze gasoline, 74, 84, 85
Brown, Bobby, 163
Bundy, Tom, 234
Burbank field, 42
Burns, Linda, 210
Bush, George W., 215, **215**
Butte, Montana, bulk station, 18, **18-19**
Butter, Don, 121, 152, 153
Buffalo Bill, 23, 27
Buffalo Creek, Colorado, service station, 37, **36-37**
Buxton, Brooks, 108

C

Caister Murdoch System, 189, 232, 233
Cambodia, 216, 237
Cambridge Energy Research Associates, 257
Canada, 148
Canada, 107
Canvey Island, England, re-gasification facility, 107
Cantu, Lisa Douglas, 174, **174**, 175
Cantu, Terry, 174
carbon fibers, 213
Carlon Products Corp., 121
Carter Oil, 100
Carter, President James Earl Jr., 152
Casbeer, Tom, 247
CATC, 104, 119
CDR 101 Flow Improver, 175
CDR 102 Flow Improver, 175, **174**
Central African Republic, 135
Chad, 135
Cherokee pipeline, 114
Chevron, 226, 227, 231
Cheyenne, Wyoming, bulk station, 19
Cities Service Co., 104, 108, 147, 155
China, 220, 221

Chevron, 152
Christ, Janice, 216
Christman, Betty, 148, **148**
Christman, C. D. "Corky," 148
Clinton, President William J., 215
coal. *See also* Consolidation Coal Company
   conversion, 122, 135, 136, 152
   industry, 121, 140, 141, 158, 160, 161
   high-sulfur, 152, 160
   mining, 22, 58, 121, 129, 136, 163
   oil, 15
Coastal Oil, 114, 160
coke, 247
   petroleum, 130, **130**, 131, 152, 173, 222
   fuel-grade, 218, 219
   needle, 222
Colonial pipeline, 114
Comar Oil Company, 59
computers
   and offshore production, 119, 153, 171
   and refraction seismography, 136, 182, 236, 239
   and research, 154
Congo
   exploration, 171, 173
   site of chimpanzee research, 248
Conly, Stan, 117
Conoco Arabia Inc., 108
Conoco capsule logo, 127, **127**, **136**, **153**, **170**, **175**, **176**, **185**, **188**, **208-209**, **222**, **237**, **255**, **263**, **264**
Conoco Challenge, 212, 213, 214, **214**
Conoco Chemicals, 121, 138, 160, 198. *See also* Vista Chemicals
Conoco European Gas Limited, 187
Conoco Global Power, 257
Conoco Global Year 2000 Program, 234
Conoco Inc. *See also* Continental Oil Company
   airplane crash, 192, 193, 194
   advertising, 175, **175**, **176**, 177, **177**, 262
   board of directors, 155, 211
   Downstream International Business Development department, 245
   employee bonus, 211, 213
   environmental record, 173, 189, 190, 192, 200, 202, 226, 227, 244, 248
   headquarters, 158, 162-163, **162-163**
   separation from DuPont, 211, 265
   IPO, 119, 211, 256-265
      celebration, 260, 262, **262**, 264, **264**
      essay contest, 260, 261, 263
      financing, 259
      roadshow, 260
   Leadership Center, 213
   marketing, 148, 160, 172, 173, 174, 175, 176, 177, 196, 198, 216, 220, 222, 235
   and the Marland estate, 62
   merger with DuPont, 155, 158, 159, 160, 163, 256
   natural gas and gas products department (NG&GP), 183, 186, 187
125th anniversary, 265

public relations, 170
restructuring, 195, 196, 197, 198, 213
retail strategy, 172, 173, 174, 175, 192, 197, 198, 216-217, 220, 222, 255
safety record, 160, 171, 180-181, 183, 226, 227
stock, 154, 155, 258, 259, 260, 264
upstream strategy, 183, 189, 195, 196, 198, 237, 240, 248, 250, 256
Conoco Indonesia Inc., 189
Conoco International Petroleum Company, 201, 202
Conoco magazine, 76, 141
Conoco Mineraloel (ConMin), 174
Conoco No.1, 110
Conoco Norway Inc., 202, 205
Conoco Show, 77
Conoco Singers, Orchestra and Players, 77
Conoco Speakers Program, 151
Conoco Tower. *See under* Houston
Conoco Travel Bureau, 78, **78**, 79, 80, 81, 83. *See also* Touraide
Conoco U.K., 234
Conoco (U.K.) Limited, 145, 170, 183, 187, 213,
Conoco World, 171, 175, 258, 265
Conorada, 108, 109
Consolidated Tank Line Company, 22, 31. *See also* Standard Oil Company
Consolidation Coal Company (Consol), 121, 122, 138, 150, 151, 160, **161**, 163, 178, 198
Constock Liquid Methane Corporation, 107
Constitution, 173
Continental, 192
Continental Carbon (Concarb), 160
Continental Oil and Transportation Company of Colorado, 20
Continental Oil and Transportation Company of California, 20
Continental Oil Company. *See also* Conoco Inc.
   advertising, 17, **17**, 31, **31**, 32, **32**, 33, 76, 77, **77**, 79, 84, **84**, 85, 122
   board of directors, 121, 129, 138
   bulk stations, 18, 19, **18-19**, 20, 21
   centennial, 150
   and competition, 22
   continental soldier logo, 36, 37, **37**, 54, 55, **55**, 265
   distribution, 18, 19, 20, 21, 31, 32, 36, 56, 73, 76, 115, 116, 117
   early products, 16, 17, 19, 20, 31, 32
   employee activities, 34, **34**, 85, 93
   employee bonus, 85
   environmental policy, 136, 152
   first overseas development, 108, 109
   first pipeline, 20
   first retail outlet, 36, 37, **36-37**
   founding, 14, 15, 16, 17, 36, 265
   headquarters, 66, 74, 91, 103, 119, 134, 138, 139
   incorporation, 54

272   125 YEARS OF ENERGY

international exploration department, 110, 129
downstream strategy, 17, **17**, 31, **31**, 32, **32**, 32, 53, 55, 56, 98, 148, 149
marketing, 20, 30, 31, 32, 53, 56, 77, 96, 114, 115, 148, 149
merger with Marland Oil, 54, 66, 73-74, 76
merger with Mutual Oil, 53, 55, 56
merger with Standard Oil, 31, 32
name change, 36
ownership, 74, 154
public relations, 142, 151
red triangle logo, 70, **70**, **71**, 73, 96, 127, **127**
research department, 101, 119
restructuring, 103, 138
retail strategy, 36, 38, 114, 115, 116, 145
safety programs, 83, **86**, 87, 89, **89**,
75th anniversary, 101, **101**,
wartime production, 92, 93, 96, **96**, **97**,
Continental Oil Producing Company, 53
Continental Oil Wheel Club, 34, 35, **35**
Continental Pipeline Company, 165
copper mining, 122, 123, 129
Council Bluffs, Iowa, bulk station, 19
Cram, Dr. Ira, 104
Crawford, John, 112, 113, 114
Cresap, West Virginia, coal-to-liquids plant, 122
Cry, Willie, 29, 39
Curtis, L. B. "Buck," 126, 128, **128**, 129, 166, 167, 169, 170, 171, 202
CUSS Group, 105, 107
CUSS I, 107
Czech Republic
cogeneration plant, 217
refineries, 217, 220
retail outlets, 198, 216, 217

D
Dahra field, 108, 109, **109**, 110
Darr Pilot Training Center, 93, 94-95
Davis, Jean "Pogo," 234
Davis, Ted, 147, 187, 234, 237
Deepwater Frontier, 253
Deepwater Pathfinder, 250, **250**, 251, **251**, **252-253**, 253, 254, 256
deepwater production, 170, 205
delayed coking process, 119, 136, 217, 219, 242, 247. *See also* coke
Delhi-Taylor Company, 187
Denmark, 174
Denver, Colorado
headquarters, 33, **33**, 34, 36, 54, 55, 66, **67**
McPhee Building, 34
marketing center, 31
refinery, 83, 103, 147
retail outlets, 175
warehouse, **31**
Derr, John, 225

detergent intermediates. *See* petrochemicals
Diassi, Frank, 121
Dietrich, Gayle, 193
Dietrich, Bill, 193
doodlebuggers, 110, **110**, 113
Dome Petroleum Limited, 153, 154
Dominion Motor Spirit, 61
Doty, Bill, 112
Douglas Creek field, 187
Douglas Oil, 115, 160, 210, 211
Dragon Trail gas processing plant, 187
drag reducers, 175
polymeric, 136
Drake, Colonel Edwin L., 14
drilling
core drilling, 58, 59, 102, 182
deepwater, 166, 170, 171, 250, 251, 252, 254, 256
horizontal drilling,
offshore, 136, 147
Dubai, 117, 119, 214
core upstream region, 196
exploration, 117
holdings, 116, 117
production, 118, 218, 219
Dubai, 116
Dubai Petroleum Company, 116, 117, 166
Dunham, Archie W., 139, 141, 150, 155, 178, 180, 192, 194, 196, 201, **201**, 205, 206, 210, 211, **211**, 212, 213, 214, 215, **215**, 231, 251, 256, 258, 260, 262, **262**, 263, **263**, 265
Dunham, Linda, 211, **211**, 263, **263**, 265
Dunlin field, 142, 145
Duplantier, Jon-Al, 119, **119**
DuPont (E. I. du Pont de Nemours & Company)
board of directors, 155, 258
Marketing Excellence Award, 175
merger transition, 160, 163
merger with Conoco Inc., 154, 155, 158, 159
plant at Orange, Texas, 187
safety record, 160
safety training, 180, **180**, **181**,
spinoff of Conoco, 210, 257, 258, 265
venture with Conoco, 153

E
Eagle, Chief White, 28, **28**, 29
Eagle, George, **28**
Earhart, Amelia, 74
East Germany, 198, **198**, 216, **216**
East Shetlands Basin, 131
Edwards, Gary, 131, 141, 150, 175, 198, 220, **220**, 221, 245, 260
Eggar, Tim, 226
Egypt, 108
Egyptian-American Oil Company, 108
E. I. du Pont de Nemours & Company. *See* DuPont
Eisler, Vic, 104-105
E. J. Nicklos, 56, **56**

electric power generation, 121, 257
Elf Aquitaine, 259
Elk Basin Petroleum Company, 53
El Paso Natural Gas, 187
energy crisis, 126, 127, 135, 137, **140**, 141, 151
gas shortage, 139, **139**, 140, 142
Energy Policy and Conservation Act, 141
Environmental Protection Agency (EPA), 136, 147
Equador, 173
E-Qual, 149
España, 131
Espinosa, Miguel "Mike," 245, 247
Es Sider pipeline, 109, 110, 116
inauguration, **108**, 109
Esso, 132
Ethyl Gasoline, 79
Europe, 148
Europe. *See* respective countries
Exel Paralubes, 222, 223, 224
exploration. *See* specific sites and projects
Exxon, 119, 189. *See also* Standard Oil of New Jersey

F
FasGas, 149, 175
Fateh field, 116, 117, 118, **118**, 126, 129
Federal Energy Administration (FEA), 141
Fisher, John, **170**
Fleming Stills, 56, **56-57**
Florence, Colorado
copper deposit, 123
refinery, 53
Forsythe, John Duncan, 64
Fowler, Sondra, 170
Fox, Ken, 193
Frantz Corporation, 53
Fuels. *See also* specific brands
alternative, 135, 152, 153
aviation, 74, 92, 93, 96, 174
gasoline, 29, 36, 37, 40, 41, 49, 53, 61, 72, 84, 114, 115, 116, 170, 173, 174, 197, 218, 219, 220, 221 (*see also* benzine)
additives, 175
diesel, 72, 218
unleaded, 175
synthetic, 136, 152, 153

G
Gabon, 171
Garner, Juanita, 213
Garvin, Barry, 245
Gault, Allen, 254, **254**
Gemini Consulting, 196
Germany. *See* West Germany; East Germany
Germ-Processed Motor Oil, 75, **75**, 76, 77, **77**, 79, 98
Gibson, Hoot, 23, 27
Gillis Gas Processing Plant and Louisiana Gas

System, 185
Glacier Pipe Line, 114
Glenn, Wayne, 83, 129
Glenrock, Wyoming, facility, 53
Global Excellence Blueprint, 195, 196
Golden Gavel, 151, **151**
Goldman, Bob, 259, 260
Goodall, Jane, 248, **248**
Gormley, Bill, 247
Gover, Bill, 175
Graham, Gerald, 224, **224**
Grasso, Richard, 263, **263**
Great Depression, 73, 74, 76, 83, 84, 85, 88, 107, 213
Great Lakes pipeline, 76, **76**, 114
Greatest Gamblers, The, 61, 65
Green Canyon Basin, 171
Gregg, Dennis, 132, 145, 173
Griffith, Dave, 240, 245, 247
Griffiths, Malcolm, 187, 226
Grivetti, E. J. "El," 142, 183
Guanare, 247
Guardian, 190, **190**, 192
Guatemala, 113
Gulf of Paria West, 247, 248
Gulf Oil, 121, 131, 132, 145
Gulf of Mexico, 119, 239, 254
exploration, 104, 137, 147, 251, 253, 254
holdings, 164, 166, 251
production, 170, 251
Gulf of Suez, 171, 173

H
Hamilton Oil, 53
Hardesty, Howard Jr., 141
Hart, Hank, 145
Heard, Paul, 247
Heidrun field
discovery, 202
platform, 202, 203, **203**, 204, **204-205**, 205
production, 231
Henkel, Tom, 212, 256, 260
Hinson, Howard, 108, 110
Hitchings, Gordon, 145
Hittle, Dick, 110, 119, 132
Holliday, Chad, 257, 258
Hollister, U. S., 22, **22**
Home Fuel Oil, 160
Horning, John, 179, 183, 202
Houston Oil & Gas Building, 91
Houston, Texas
Conoco Tower (Greenway Plaza), 138, 139, **139**, 163
headquarters
Sterling Building, 91, 103, 119
Houston complex, 162, **162-163**, 163
Howard-Glasscock field, 60-61
Howerton, Billy Joe, 185
Howerton, Gene, 185
Howerton, Levi, 185
Howerton, Pat, 185, **185**

INDEX 273

Howerton, V. C. "Pat," 185, **185**
Howerton, Virgil, 185
Hudson's Bay Company, 61
Hudson's Bay Oil and Gas Company (H-BOG), 107, 138, 153, 154
Humber, 131
Humber refinery, 130, **130**, 131, 142, 145, 173, 185, **185**, 220, 222
Humble Oil Company, 100
Hungary, 198, 216
Hutton field, 166, 169, 171, 226
    platform, 145, 167, **167**, **168**, 169, **169**,
    christening, 169
    commemorative postage stamp, 169, **169**, 170
Hydroclear, 223, 224, **224**

I

Independence, 148, 249, **249**
India, 220, 221
Indonesia
    core upstream region, 196, 137
    development, 189, 194, 234
    exploration, 119, 137, 147, 171, 189, 225,
    natural gas holdings, 189, 225
    production, 218, 219, 234
    retail outlets, 220, 221
Ingleside Cogeneration, 257, **257**
Interconnector pipeline, 233, 234
Iran
    holdings, 151, 214
    negotiations, 214, 215
    sanctions against, 214, 215
Iranian International consortium (IRICON), 151
Ireland
    aviation market, 174
    liquid petroleum gas (LPG) market, 174
    marine market, 174
Irian Jaya. *See* Indonesia
Israel, Bob, 256
Italia, 117
Italy
    refinery, 115
    retail outlets, 145, 172, 173

J

James, Steve, 193
Jefferson, Edward G., 155, 158, 160, 163
Jet, 131
Jet Petroleum Limited
    acquisition, 114, **115**, 116
    retail outlets, 114, **114**, 172, 173, 174, 175, 197, 198, **198**, 198, 216, **216**, 221, **221**
Jiffy stores, 174, 175
Johnson, Mike, 205
Johnston, Gary, 193
Jolliet field, 170, **170**, 171
    platform, 171, **171**, 251
Jones, Wanda Lee Fisher, 88

Jose, Venezuela, upgrader facility, 241, 242, **242**, **243**, 247
Joy, Maria, 115
Jurenev, Serge, 101
J. W. Green Mercantile Co., **36-37**, 37

K

Kah, Marianne, 225
Karlsruhe, West Germany, refinery, 115
Kay County Gas Company, 29, 40
Kayo Oil, (Kendall Oil), 114, 160
Kayo retail outlets, 114, 115, 160, 197, **197**
Kem, David, 173, 174, 217
Kemp, John, 205, 260
Kendall Oil. *See* Kayo Oil
Kenney-Cleary Company, 42
Kenney, Franklin, 23, 27, 42
Keoughan-Hurst Drilling Co., 53
Keoughan, S. H., 53, 56
Kerosene, 16, 17, 32, 34, 36
Khazzan, 126, 128, **128**, 129
Kim, J. W., **253**
Kim, Mrs. J. W., **235**
Knowles, Ruth Sheldon, 61, 65
Knudsen, Tom, 187
Kotarski, Joe 121
Kotter field, 166
Kralupy, Czech Republic, refinery, 217, 220
Krasts, Aviars "Ike," 122, 160, 163

L

Lafferrandre, Bill, 216
Lake Charles, Louisiana
    coker, 245
    pipeline, 88
    Port of, 148, 192
    power plant, 152
    refinery, 88, 92, 103, 147, 149, 153, 173, 222, **222-223**, 223, 224, 245, 247, 254
    laboratory, 120, **120**
    liquefaction plant, 107
lamp oil, 16
Lanagan, Missouri, service station, 41
Lashbrooke, Paul, 218, 220
Lauren, P. K., 145
Lavoisier Medal for Technical Achievement, 112, 113, 167
Leduc field, 107
Lee, Bob, 210
Lee, Brooke, 193
Lee, Colin, 193, 245
Lesser, Marvin, 132
Lexington, 171
Libya, 179, 183
    holdings in, 108, **108**, 109, **109**, 110, 183
Libya, 116
Life and Death of an Oilman, 39, 42, 61
Lincolnshire Offshore Gas Gathering System (LOGGS), 188, 189, 232, 233
Lindbergh, Charles A., 74, **74**

liquefied petroleum gas (LPG), 106, 107, 218, 219, 247
Liquid Flow Improver, 174
Little Will can, 17, **17**
Litvinov, Czech Republic, refinery, 217, 220
Liu, Daisy, 216
Lloyd, Paul, **253**,
Lloyd, Penny, **253**
Lobo trend, 236-239, **239**
    drilling, 236, **236**, 238, **238**
    pipelines, 238, **238**
Logger field, 166
Louisiana, 137, **137**, 148, 149, **149**
lubricants, 222, 223, 224, 256. *See also specific brands*
    grease, 16, 41
    household products, 98, **98**
    motor oil, 53, 74, 75, 76, 77, **77**, 79, 88, 98, 101, 102, 114, 119, 172
    synthetic, 119
Lukoil, 251
Luxembourg, 114, 145
Lyell field, 226

M

Mablethorpe, England, 188
Madagascar, 135
Malaysia, 256
    refining, 207, 217
    retail outlets, 220, 221
Marathon Oil, 108, 154, 155
Marland, Ernest Whitworth, 14, 22, 23, **23**, 25, 27, 28, 29, 40, 43, **43**, 50, 60, **60**, 61, **61**, 64, 65, **65**, 70, 102, 265
    benefactor, 38, 46, 61, 62
    employer, 46
    geologist, 27, 39, 50, 58, 59
    politician, 61, 65, 66, **66**, 67
Marland estate, 61, 62, **62-63**, 61, 67
Marland, Lydie Roberts, 61, 62, **62**, 65, 67
Marland, Mary Virginia, 62
Marland Axle Grease, **46**
Marland Oil Company, 42, 50. *See also* Marland Refining Company; 101 Ranch Oil Company
    advertising, 29, **29**
    board of directors, 60, 65
    debt, 59, 61, 65, 73, 74
    employee benefits, 46
    establishment, 42
    exploration, 50, 59, 98, 100, 119
    headquarters, 64, **64**, 65
    international markets, 50, 59, 61
    marketing, 28, 40, 59, 73, 74
    merger with Continental Oil Company, 54, 66, 67, 73-74, 76
    red triangle logo, **29**, 42, 46, **46**, **47**, **48**, 67, 70
    refining, 39, 42, 50, 58, 70, 73, 74
    research division, 58, **58**, 59, 101
    retail outlets, 46, **46**, 48, 50, **50-51**, 60

Marland Refining Company, 39
    cup grease, 41, **41**
    gas-testing machine, 41
    products, 40
    retail outlets, 40, **40-41**, 41
Marr, Tom, 145
Marshall, Keely, 71
Marzuola, Randol, 213
Matthews, John Joseph, 39, 42, 61
Mauritania, 113
Mayer, Mike, 196, 198, 256-257
McAteer, George, 108, 113
McCall, Bruce, 149, 152
McCollum, Leonard F., 91, 99, 100, **100**, 101, 102, 103, 104, 108, 119, 120, 123, 129, 134, 135, 179, 265
McFadden, W. H., 29
McGeachie, Dan, 169-170
McKee, Ann, **253**
McKee, Rob, 158, 165, 179, 183, 196, 198, 226, 231, 234, 237, 239, 245, 247, 248, **253**, 254, 256, 260
McKinsey & Company, 257
McKonkey, 222, **222**
McLean, John Godfrey, 119, 126, 129, 130, 131, 135, **135**, 136, 137, 179
Melaka, Malaysia, refinery, 217, **217**, 218, **218-219**, 219
Melicharova, Lucie, **261**, 263
Merritt Oil, 53
Methane Pioneer, 107
Methanol, 152
Meyers, Brent, 171
Milan, Italy, oil distribution terminal, 115
Miller, Colonel George Washington, 25, 28
Miller family, 23, 25, 27, 28, 29
Miller field, 166, 194
Miller, George L., 23, **24**, 27, 29
Miller, Joe C., 23, **24**
Miller, Dr. Walter, 87
Miller, Zachary T., 23, **24**
Milne Point field, 164, **164-165**, 165, 166
Missippi Canyon, 164, 166
Missoula, Montana, bulk station, 20, **20**
Mitchell, Claudia, 79, 83, 84
Mix, Tom, 23, 27
Mobil, 119, 132, 154, 155
Mock, Mildred, 83
Moran, Dan, 65, 70, 71, 73, 74, 78, 82, **82**, 88, **88**, 96, 99, 101, 213
Morgan, Franklin, 32
Morgan, J. P. Jr., 60, 65, 74, 82
Morris, Mike, 131, 163
Morrow, John, 74, 103
Mueller, Tom, 222
Murchison field, 142, 143, 226
    development, 145
    platform, 142, **142-143**, 143, **144**,
Murphy, John, 107
Mutual Oil Company, 53, 56
Myers, Jim, 193, 245
Myers, Linda, 193

N

Nabors, Hal, 166
Nalkylene Alkylate, 120, 121
naptha, 42, 218, 219, 247
National Coal Board of Britain (NCB), 131, 132
National Gas Company of Trinidad and Tobago, 189, 225
National Iranian Oil Company (NIOC), 214, 215
National Petroleum Council (NPC), 134
    Committee on the U.S. Energy Outlook, 134, 135
National Safety Council, 87
    Trophy Award, **86**, 87
natural gas and gas liquids, 106, 107, 132, 145, 166, 183, 186, 214, 227, 239, 240
Nenets, 201
Newell, Ozzie, 173
New York City, **261**
    headquarters, 119, 134
New York Stock Exchange, 74, **208-209**, 211, 256, 260, 262, **262**, 263, **263**, 264, 265
New Zealand, 254
Neylon, Bian, 214
Nicandros, Constantine S. "Dino," 104, 119, 123, 126, 155, 158, 160, 163, 178, 179, 183, **188**, 189, 190, 192, 193, 194, 196, 198, 199, **199**, 202, 205, 206, 207, **207**, 214, 245
Nigeria
    exploration, 199, 254
    production, 249
Nixon, President Richard M., 141
Nokes, Jim, 196, 224, 260
Norske Conoco, 145
North America
    core upstream region, 196
    downstream structure, 197, 198
    exploration, 114, 137, 146, 164, 251
    natural gas holdings, 187, 236, 237, 238, 239, 240
    production, 146, 186, 195, **195**
    retail outlets, 76, **84**, **90-91**, 91, 98, 114, 115, 149, 175, 197, 198, 254
Northern Territories (Russia), 251
North Rockall Trough, 249
North Sea
    core upstream region, 196, 202, 248, 256
    development, 131, 132, **132**, 142, 145, 166, 187, 188, 194, 202, 225, 227
    exploration, 119, 129, 136, 137, 142, 166, 225, 226
    explosion, 183
    natural gas holdings, 187, 225, 226, 227, 231, 232, 233, 248
    Southern Basin, 129, 166, 187, 188
Norway
    development, 202, 203, 204, 205
    exploration, 202, 249, 254
    holdings, 132
    natural gas holdings, 202

    retail outlets, 198
Norwegian Sea, 202, 203, 231
N-tane, **73**
Nth Motor Oil, 88

O

Oasis Oil Company, 109, 110, 179
Oasis Petroleum, 160
Occidental Petroleum Corporation, 183, 257
Offshore Technology Conference, 170
Ogden, Utah
    headquarters, 15
    bulk station, 19
Ogren, John, 183, **188**
Ohio Oil Co., 108, 109
Oshlo, Rick, 189
oil
    extra-heavy crude, 152, 218, 241, 242, 245, 246, 247
    heating, 174
    hydrocracked base, 223, 224
    motor (*See under* lubricants)
    synthetic crude, 247
oil embargo, 126, 127, 134, 137, 138, 140; *See also* energy crisis
OK Coop, 174
Oklahoma, 148
Oklahoma, state of, 62
    University of Oklahoma, 38, 210
Oldfield, Sue, 184, **184**
Oldfield, William, 184, **184**
101 Ranch, 23, 24, 25, **24-25**, 29, 39
    Real Wild West Show, 23, 25, 26, **26**, 27
101 Ranch Oil Company, 29
    first oil, 29
    founding, 27
    pipeline to Tonkawa, 29
    production, 39
Orange, Texas
    DuPont plant, 187
Organization of Petroleum Exporting Countries (OPEC), 127, 134, 137, 138, 139, 179, **179**
Orinoco belt, 241, 242, 245, 246
Oryx, 226
Osage Indian tribe, 40
Oklahoma City, Oklahoma, service station, **48**, 49

P

Pacesetter II, 147
Pan West Engineers and Constructors, 189, 225
Paramount City, California, refinery, 147, 160
Parsons, Ann, 193
Patriot, 189
Pawhuksa, Oklahoma, service station, 42
Parker, Dennis, 192
Pearce, Mary Ann, 234
Pennzoil, 222, 223
Perlitz, Charles, 121

Permian Basin, 147
Perrine, Irving, 39
Persian Gulf War, 194
Pertamina, 189, 234, 235
petrochemicals, 42, 92, 96, 101, 103, 120, 121, 160
Petroleos de Venezuela (PDVSA), 241, 245, 247, 248
petroleum industry (U.S.)
    birth of, 14
    declining growth, 214
    early production, 16, **16**
    early distribution, 18
    and the energy crisis, 127, 139
    regulation of, 136, 140, 141, 151, 165, 189
    and U.S. foreign policy, 214, 215, 216
Petro-Lewis Corporation, 160
Petronas, 217, 218
Petrozuata, 202, 241, 244, 245
    drilling, 246, **246**, 247
    pipeline, **240-241**, 242,
    project financing, 245, 247
Phoenix Park Gas Processors, Limited, 189, 225, **225**
Pickett, Bill, 23, 27, **27**
Pioneer, 192
Pioneer pipeline, 114
Pioneer Woman, **62**, 62, 65
Pitcher, Max, 199, 201
Platte Pipe Line Company, 114
Pocahontas, Virginia, 187
Point Lisas, Trinidad, gas plant, 225, **225**
Poffenbarger, Gary, 247
Poland, 198, 216
Polar Lights, 199, **199**, **200**, 201, 202, 251
    negotiations, 199, 201
Ponca City, Oklahoma, 23, 27, 38, **38**, 39, 47, 50, **56**, 61, 62, 72, **72**,
    compound and packaging plant, **74**
    headquarters, 64, 74, 91, 103
    propane cave, 106, **106-107**
    refinery, 39, 42, 58, 70, **70**, 75, 92, 99, 119, 147, 148
    refinery expansion, 59, **59**, 103
    research laboratory, 101, 102, **102**, 110, 153, **153**, 224
    Sequoia refinery, 147
Ponca Indian tribe, 23, 25, 27, 29, **38**. *See also* Eagle, Chief White
    land, 27, **27**, 28, 29, 39
Porcupine Basin, 249
Porter, Les, 234
Potsdam, Germany
    office, 198, **198**
    service station, 198, **198**
Potter, W. H. "Billy," 19, 34, **34**
Powerscrub, 175
Prieur, Jean-Marie, 145, **145**
Prince William Sound spill, 189, 190
Prize, The, 139
Profit Taker, 87
Project 2000 Blueprint, 195, 196

ProJET, 220
Prudential Refining Company, 73, 74
Punches, Max, 140
Pure Oil Company, 210

Q

Qatar
    exploration in, 135, 137
Queensland, Australia
    exploration in 113

R

R&B Falcon Corporation, 192, 254
Ramshaw, Roger, **188**
Randgrid, 192
Red Triangle, 110
refraction seismography, 58, 59, 104, **104**, **105**, 112, **112-113**, 113, 182
    computers and, 136, 182, 236, 239
    crews, 110, **111**, 112, 113 (see also doodlebuggers)
Reagan, President Ronald, 165
Richfield, 108
Riley, Charles David, 185, **185**
Riley, Charles William, 185, **185**
Roberts, Andrew, 213, **213**
Roberts, George, 62
Roberts, Lydie. *See* Marland, Lydie Roberts
Robertson, Don, 233
Rocconi, Mike, 198
Roche, Kevin, 162
Rockefeller, John D., 19, 22, 31, 39, 53
Rodriguez, Jose, 145
Rogers, Ginger, 79
Rogers, Will, 67
Roley, Roy, 116
Rooney, Mike, 260, 261, **261**
Rostock, Germany, retail outlet, 198, 216, **216**
Royal Dutch Shell, 40, 59
Russia, 256
    development, 200, 202
    exploration, 199, 201
Ryman, Lloyd, 131, 132

S

Saga, 132
Sager, Harry, 117, 142, 166, 169
Salupa Refining Company, 53
Samsung Heavy Industries, 190, **191**, 253, 254
San Jacinto, 151
San Juan Basin, 186, 187, 189
San Juan, New Mexico
    gas-processing plant, 186, **186-187**, 188
    holdings, 187
Santa Maria, California, refinery, 147
Sarrah, Dr. Marion "Slim," 101
Schwartz, Samuel, 126
Schwartzman, David, 141
Seagram Company Limited, The, 154, 155
Sealand Petroleum Company, Limited, 50, 61

Seal Beach field, 61
Seaway pipeline 148
Seca
    acquisition, 114
    retail outlets, 114, 115, 116, 172, **172-173**, 173
Seca, 116
Securities and Exchange Commission (SEC), 259
Seismography. *See* refraction seismography; Vibroseis
Sellers, Jen, 55, 56
Sembawang Engineering and Construction, 234
Senegal
    exploration in, 113
Senseman, Irene B., 96
Sentinel, 173
Sequoia refinery. *See under* Ponca City
Semenovich, Vladimir, 201
Shafranik, Yuri, 202
shale oil, 152
Shell, 104, 132, 189, 217
Shetland Islands, 142, 145, 166, 248. *See also* East Shetlands Basin
Sigler, Tom, 136
Simpson, Dee, 216
Singapore, 234, 235, **235**, 237
    gas processing facility, 234
    subsea pipeline to, 234, 235
Sirri field, 214, 215
Slavin, Vincent, 258
Slovakia, 216, 217
Smith, Emerson G., 33, 36
Smith, Leland, 27, 29, 71
Snell, Ed, 55
Somaliland
    exploration in, 113
Sopi, 115,
    retail outlets, 114, 115, 116, 173
Sopi, 116
South Ponca field, **42**, 43
Spain, 198
specialty chemicals. *See* petrochemicals
Stafford, F. D., 22
Stamford, Connecticut, headquarters, 134
Stanat, Jeff, 247
Standard Oil Company, 22, 31, 34, 36, 39, 40, 54, 158
Standard Oil of New Jersey, 59
Statfjord field, 132, 133, **133**, 142, 145, 232, 233
Statfjord C field, 166
Statoil, 132, 205, 217, 218
Steinfort, "Red," **86**, 87
Stewart, Ray, 77
Stinson, Mike, 107, 215, 240, 245, 248
Strong, C. E., 55
Stolt Migration Process, 113
Stolt, Robert, 113
Strickland, John W., 119, 147
Submarex, 107
SubSea MudLift Drilling Project, 254

sulfur, 247
Superion, 104
Super Motor Oil, 101, 102
Sweden
    home heating oil, 174
    retail outlets, 174
Switzerland, 174

T
Taiwan, 216
Target 2000 (downstream), 196
Tarkington, Andrew W., 121, 126, 129, **129**, 136
tension-leg, 128, **128**, 166, **166**, 167, **167**
    platform (TLP), 145, 166, 167, **167**, **168**, 169, **169**, 170, 251
    concrete platform, 202, 203, **203**, 204, **204-205**, 205
    well platform (TLWP), 171, **171**, 251
Tetlow, Jeff, 226
Texaco, 119, 121, 131, 132
Texas, 148
Texas, 236-239
Texas City, Texas, 59
Texhoma Oil and Refining Company, 56
Thailand
    exploration, 237
    retail outlets, 198, 220, **220**, 221, **221**
Thatcher, Prime Minister Margaret, 169, 188, **188**, 189
Theddlethorpe, England, gas terminal, 131, 189, 230, 233
Thistle field, 142, 145
Thomas, Bill, 99
Thorpe, Donna, 184, **184**
Thorpe, Jack, 184, **184**
Thorpe, Samuel, 184
3-D seismic imaging, 236, 239
Thorpe, Sidney "Pug," 184, **184**
Three Sands field, 43, **44-45**
Tidewater Oil, 104
Tilford, Henry Morgan, 32, 33, **33**, 34
Timan Pechora Basin, 201
Tjeldbergodden, Norway, methanol plant, 205, 231
Tonkawa field, 43, 47
Total, 215
Touraide, 81
Trans-Alaska pipeline, 164, 165
Trans-Alpine pipeline, 115
transportation
    horse-drawn tank wagons, 21, **21**
    motorized tank cars, 36, **36**
    pipelines, 76, **76**, 174
    tankers, 56, 107, 110, 116, 117, **124-125**, 131, 147, 148, 149, 173, 247, 251
        double-hulled, 189, 190, **190**, 192
    tank rail cars, 17, 70, 71
    tank trucks, **72**, 73, 76
Travel Bureau. *See* Conoco Travel Bureau

Trinidad and Tobago
    exploration, 251
    natural gas plant, 189
Trinidad, Colorado, compounding plant, 20, 20, 22, **22**
Troll field, 232, **232**, 233
Tubby and Red, **86**, 87
Tunisia
    exploration, 113
Turbo 10 W-30 Motor Oil, 172, **172**
Turkey
    exploration, 113
    retail outlets, 216, 217
Turvey, Bob, 131

U
Udang field, 147
Udang Natuna, 147
Ukpokiti field, 199, 249
Union, 104
United Arab Emirates, 119
United Kingdom,
    aviation market, 174
    Department of Trade and Industry (DTI), 226
    liquid petroleum gas (LPG) market, 174
    marine market, 174
    natural gas holdings, 187, 188, 225
    natural gas market, 189, 225, 230
    retail outlets, 114, **114**, 116, 145, 172, 173
United Oil Company, 53
Union Stock Yard Company, 107
Unocal, 154, 155
Unruh, Don, 245
Uranium mining, 122, 123, 129, 136
Uranium Exploration Group, 123
Ursa field platform, 171, 251, 254

V
Valdes, Antonio, 189, 254
Valdez, 189, 190
Valiant field (North), 187, 188
Valiant field (South), 187, 188
Vanguard field, 187, 188
Van Wageningen, Henry, 247
"V" fields, 187, 188, 232, 233
    inauguration, 188, **188**, 189
Venezuela
    core upstream region, 248, 256
    exploration, 113, 207, 240, 245, 246, 247
Vibroseis, 101, 102, 112, **112-113**, 113-114
Vicrachat, S, 220
Victor field, 166, 187, 188, 232, 233
Vietnam
    exploration, 216, 237
Viking field, 132, **132**, 145, 232, 233
Viking Transportation System, 189, 230
Vista Chemicals, 160
Vøring Basin, 249
Vulcan field, 187, 188

W
Wagstaff, Mike, 256
Walmsley, Les, 185, **185**
Walmsley, Lisa, 185, **185**
Warwick, England, offices, 116
Watkins, George, 234
Webster, Kieth, 249
Wentworth, Ralph, 77
Wentz, Lew, 29
Wertheim Schroder & Co., 196, 256
Western Oil and Fuel, 114
    retail outlets, 114, 115
    convenience stores, 115, 160
Western Oil Fields Corporation, 53
West Germany, 115, 145, 172, 173, 174
West Natuna Sea, 234, 235, 237
Whitman, John, 147
Wibipriatno, Titie, 260, **261**
Wilbur, Herbert "Tubby," **86**, 87
Williamson, Liz, 212
Willie Cry well, 29, 39, 43, **43**
Wilmington, Delaware, headquarters, 158
Wilson, B. G., 32, 55
Wilson, Edward T., 34, 36, 53
Wing, Dr. Richard, 145
Wining, Cathy, 179
Witco Chemicals, 160
Wong, Sing, 216
Woolard, Ed, 210
World Energy Council (WEC), 215
World War I, 40
World War II, 73, 92, 93, 96
Wrenshall, Minnesota, refinery, 114, 147, 160

Y
Yamani, Sheik, 179, **179**
Yellowstone National Park, **52**, 53
    contract with Continental Oil, 53
Yellowstone pipeline, 114
Yeltsin, President Boris, 199, **199**
Yergin, Daniel, 139
York, James, 73

Z
Zama, Bob, 84
Zebrugge, Belgium, 233
Zemella, Jorge, 245
Zuata field, 241, 246, **246**, 247